FROM
LYSANDER
TO
LIGHTNING

TEDDY PETTER,
AIRCRAFT DESIGNER

GLYN DAVIES

To Adam,

for Glyn.

The
History
Press

First published 2014

The History Press
The Mill, Brimscombe Port
Stroud, Gloucestershire, GL5 2QG
www.thehistorypress.co.uk

British Library Cataloguing in Publication Data.
A catalogue record for this book is available from the British Library.

ISBN 978 0 7524 9211 7

Typesetting and origination by The History Press
Printed in Great Britain

CONTENTS

FOREWORD

Few interests are as all consuming as aviation. To be able to fly has been the dream of mankind for centuries. It has been the stuff of folklore, the pursuit of the brave, and the goal of the engineer.

Warfare, travel, exploration and sport have all been affected. The pursuit of flight dominated the twentieth century to shape our modern world.

The story of aviation has largely centred on the deeds of the heroic pilots, but the miracle of flight is primarily due to the work of engineers. Only a select few aircraft designers have come to the fore to become household names, like R.J. Mitchell, Sydney Camm or Barnes Wallis.

One name that has remained in obscurity for some time is that of William Edward Willoughby Petter.

William Edward Willoughby Petter was the son of Sir Ernest Petter, founder and chairman of Westland. He was brought into the company by his father, and his rapid rise to a senior position was an unbridled display of nepotism. However, 'Teddy' Petter, as he was known, did not need any help to stand out as a talented engineer and his name is associated with several famous aircraft such as Lysander, Whirlwind, Welkin, Canberra, Lightning and Gnat.

He was intensely shy and reserved, totally ill at ease when in company or the centre of attention. He was camera shy, appeared to be incapable of giving praise and did not suffer anyone he deemed to be a fool, and in his opinion there were many, including some of his colleagues, who fell into that category. In 2015, Westland will celebrate 100 years of continuous operation as an aircraft company, and it was felt that this would be an appropriate time to produce a biography for Teddy Petter.

Professor Glyn Davies has indeed filled the gap; not only does he discuss the technicalities of aircraft design in some detail, but he also gives a deeper insight into the character of 'a difficult man to know and a hard man to work with'.

I feel that some comment must be made concerning Petter's involvement in the Lightning. He was primarily involved during the conceptual stage that resulted in its unique configuration, the awesome operational fighter came later.

This book is not simply a catalogue of the aircraft design and problems solved, it also relates a tragic human story. In later life Petter's wife contracted Parkinson's disease and, in the course of his efforts to find a cure, this strong-minded, intelligent man fell into the clutches of a medical religious charlatan. I suspect that by the time readers have finished with this book there will not be a dry eye in the house.

At last the Petter story has been told, and Teddy Petter can take the place in history he deserves

David Gibbings MBE, FRAeS, 2014
Retired Flight Test Engineer and WESTLAND Historian

PREFACE

In 1955 I started work in the advanced projects office at the (then) Bristol Aeroplane Company, working on the research aircraft type 188 (the all-steel thin-winged Canberra!) and the type 223 (to become Concorde). At this time the name W.E.W. Petter had already become legendary, and he was known to be a creative genius but with a strange, eccentric character. In 1959 I emigrated to Sydney for six years, returning as an academic to the department of Aeronautics at Imperial College, London. During my forty-five years at Imperial, I worked on many contracts with Industry but the name of 'Teddy' Petter never cropped up. His name was nowhere near as famous to the public as were his aircraft. The Lysander became known for its adventurous trips to occupied France, to deliver and take back intelligence agents; the Canberra was the first English jet bomber and was continually in the newspapers when it shattered numerous world records for long-distance flights to most countries around the globe; and the Gnat became famous when it was adopted by the RAF as its aerobatic display team, the Red Arrows.

These aircraft were all radical designs, being totally new in concept, and nothing like anything designed before or after. They were also commercially successful. Uniquely they were designed by one man at different companies as he moved from Westland at Yeovil to English Electric at Warton, and finally to Follands at Hamble. This man was clearly a radical thinker with a character to match. He was intolerant of second-class engineers, of managers, and of Government officials, so much so that eventually he had to move on to another company and leave his frustrations behind. His move to Follands at Hamble was his final one, where he became the Managing Director and subservient only to the Board of Directors. Only when Follands were taken over by Hawker Siddeley did he resign and leave England never to return. On retiring myself I did think that there was here a story to be told.

This book tries to capture the personality and ability of an innovative designer, but also why his aircraft were so original in their overall design configuration, their aerodynamics, their structures and their control systems. Some of these features are therefore described in depth, referring to detailed appendices where necessary. Petter's designs are compared with other international contemporary aircraft, and shown to be well ahead of their time.

I did wonder why no biography had ever been written about this strange man, and discovered that in 1991 a book had been compiled consisting of articles from a dozen of his colleagues .This was never published, so I decided to write this biography using much of the material from the unpublished work. I hope I have done justice to a man who clearly inspired all who worked for him in industry, and yet whose final days were very sad, to say the least.

I should like to acknowledge the exceptional help given by Dave Gibbings of Westland Heritage, without whom I would not have taken on the task of this biography, and who provided much of the information and illustrations of Petter's designs at Yeovil. Thanks also to Dennis Leyland of BAESystems Heritage for information and photographs of the aircraft at English Electric, Warton. It is not possible directly to thank Robert Page, Roy Fowler, and Adrian Page, as they are sadly no longer with us, but they compiled a biography of Petter in the form of contributions from twelve of his co-workers and others. It was never published, but is widely cited in this book. I am also most grateful to English Electric colleagues of Petter, Frank Roe (Aerodynamics, and former MD) and Alan Constantine (former Assistant Designer), for confirming or correcting parts of the English Electric chapters. I received much help from libraries, including the British Library, but particularly from Brian Riddle at the National Aerospace Library at Farnborough. I would like to thank Andrew Doyle, Head of Content at Flightgobal for the many cutaway drawings appearing in past *Flight* magazines. Hugh Evans should be thanked for acting as an intended reader of each chapter.

I must thank all my colleagues at Imperial College, and countless students and postgraduates who kept alive my interests in aeronautical research and design.

Finally I am most grateful to The History Press for taking a gamble. Biographies of the famous and gifted are common. Books about particular military aircraft, in great depth, including their operational history, are also common. To attempt both at once is not so common.

Glyn Davies, 2014

INTRODUCTION

The names of several Second World War aircraft designers are deservedly well known. R.J. Mitchell (born 1895) is for the Spitfire; Sydney Camm (b. 1893) has the Hurricane; and Roy Chadwick (b. 1893) designed the Lancaster. The name of 'Teddy' Petter (b. 1908) has become almost unknown, and yet his aircraft are nowadays legendary, such as the Lysander, Canberra, and Gnat. Additionally, Petter designed these aircraft at three different companies, Westland, English Electric, and Folland. Another unique feature of these aircraft was their radical nature; nothing like them has been designed before or after.

The story of this designer is of a rare individual with singular talents, and an inability to suffer fools and poor management. He was supportive of his good engineering teams, but his intolerance of a hostile management led to his resignation from three manufacturers, even though he was Managing Director of his last company, Folland.

This biography tries to capture the mercurial nature of a basically shy character, and also the radical nature of his designs. These are briefly compared with the worldwide aircraft at the time, in terms of their details and performance. This is naturally a technical judgement, backed by illustrations and diagrams, so appendices on aerodynamics and structures are included, should they be needed. It is hoped that this biography will lead to recognition of Teddy Petter amongst the truly great aircraft designers.

1

FOUR GENERATIONS OF PETTER ENGINEERS, 1840–1935

William Edward Willoughby (Teddy) Petter seems to have inherited his flair for engineering design, and spotting an opportunity, from several previous generations of his family, all from the West Country. In the 1840s John Petter (Teddy's great-grandfather), an ironmonger at Barnstaple, had built up a considerable fortune, enough to buy for his son James Beazley Petter a business, Iron Mongers of Yeovil. James Petter (Teddy's grandfather) was clearly a competent and energetic businessman, and was soon able to buy outright the Yeovil Foundry and Engineering Works. He invented a high-quality open-fire grate called the Nautilus in several versions 'for the study, the dining room and the boudoir'.[2] His works employed some forty men making castings and repairing agricultural machinery. The Nautilus grates were to become famous after being selected by Queen Victoria for installation in the fireplaces of Balmoral Castle and Osborne House in the Isle of Wight.[1] His business did not make James a wealthy man, mostly because he had fifteen children to support, a large family even for the Victorian era. The third and fourth of these children were the twins Percival Waddams and Ernest Willoughby, the latter of whom was Teddy Petter's father.

We have a clear account of Percival's life written by himself,[2] and it is obvious that this large family was brought up with strict Christian beliefs. His daughter has said that Percival's faith enabled him to accept the early death of his two sons.[2] At the age of only 20 Percival took over as manager of the engineering works in 1893. He had inherited the inventive streak and, together with B.J. Jacobs, the foreman of the foundry, he designed and built in 1894 'the Yeovil Engine', a high-speed steam engine. More importantly, a year later he designed and built a small 2.5hp oil engine for agricultural use. It was immediately successful and the business expanded so swiftly that by 1904 over a thousand engines had been sold, ranging from 1hp to 30hp in size.

Teddy's father, Ernest, seemed to think that his own talents were in enterprise and business, rather than sharing his brother's engineering skills. He worked hard at becoming part of the establishment, and spent more time in London than Yeovil. In fact by 1924 he had stood for Parliament twice (unsuccessfully) and had become chairman of the British Engineers Association. He was given the task of organising the engineering section of the British Empire Exhibition at Wembley, for which the king conferred upon him a knighthood in 1925.

By 1901 the business was growing so rapidly that James Petter could not cope and eventually had a nervous breakdown. Ernest consequently joined his brother and together they bought the business off their father. With much effort they succeeded in raising £3,850 from friends and investors. They made Jacobs the chief engineer, a position he held until his death in 1936. By 1908 the sale of their engines had increased, with a very large number of orders from Russia, where they preferred a two-stroke engine

The Petter twins, Percival and Ernest. Teddy's father Sir Ernest is seated on the right.

to the older four-stroke versions. In 1911 the company was awarded the grand prix at the Milan International Exhibition for their machines, which now ranged from 70 to 200hp By now the Nautilus works employed 500 people and some 1,500 engines were produced annually. A new foundry was needed and built, completed in 1913, at which point it was one of the largest in Britain. Harald Penrose (who became their first test pilot) later recalled that Percival Petter, his wife and two daughters were present when the first turfs for the new foundry were cut at a site west of Yeovil. Mrs Petter consequently chose the name 'Westland' for the proposed factory and planned garden village.

In 1915 Lloyd George made a speech in Parliament in which he frankly exposed the inadequacy and unsuitability of the munitions available for continuing the war. A board meeting of Petters Ltd passed a resolution placing at the disposal of the government the whole of the new factory to make anything the government might call for. The War Office did not respond immediately, but the Admiralty asked for a conference, so Ernest and Percival went to London to meet five gentlemen, three of whom were Lords of the Admiralty, who stated that the great need of the Navy was for seaplanes, and they asked whether Westland were willing to make them.

The brothers explained that their 'experience and the factory were not exactly in line with these requirements but we were willing to attempt anything which would help the country'. The sea lords replied, 'Good. You are the fellows we want: we will send you the drawings and give you all the help we can. Get on with it.' Percy recalled that they sent representatives to Short Brothers at Rochester, Kent, to find out what was expected.[2] 'I must confess that my heart nearly failed me when I saw the nature of the project.' However, he then remembered a Robert Arthur Bruce, whom he had interviewed a year earlier, and who was then manager of the British and Colonial Aeroplane Company at Bristol. He was now the resident Admiralty inspector to the Sopwith Aviation Company at Kingston-upon-Thames. The Admiralty agreed to release him to knock the new factory into shape and realise the output of this manufacturing centre at Yeovil. The work considered for this new project needed a separate title from that of the established oil engine manufacturing company. Hence, although wholly owned by Petters Ltd, it would be operated as a self-supporting Westland Aircraft works, and Ernest Petter would be chairman. The split between the brothers was now official. Percival, the engineer, would look after the Petter engines and Ernest, the businessman, would guide Westland Aircraft Ltd.

One of the very first acts of this company was to recruit a young man named Arthur Davenport, who was appointed as chief draughtsman, and charged with a stay at Short Brothers. There he made necessary production drawings of the Short 184 seaplanes, for which an order for twelve had been placed. The fourth aircraft to be built at Yeovil saw action during the Battle of Jutland, operating from the seaplane carrier *Engadine*. It successfully reported by radio the movements of German fleet and consequently gave much confidence to the workforce at Yeovil that they were capable of building aircraft as good as those from any other source. Robert Bruce had proved to be a giant in capability, creating a large significant aircraft works from the small shops he had inherited. During the war he was instrumental in seeing over 1,000 biplanes built at Yeovil. They included sub-contracts for Sopwith landplanes, de Havilland DH4 Bombers and the DH9 two-seater bomber. For his wartime work Bruce was awarded the OBE.

After the war the aircraft manufacturers knew there would be no market for military aircraft, but assumed that civil aircraft would no longer be the toys of the wealthy sportsman. However, the public was not in the least air-minded and over a decade would pass before a market emerged, in spite of the Department of Civil Aviation funding competitions for small and large aircraft and seaplanes. Westland had no crystal ball, so in 1922, when the Air Council announced a light aircraft competition, Robert Bruce submitted a biplane and Arthur Davenport a monoplane. Both prototypes were built and flew in 1924. Thirty of the (Widgeon) monoplanes were eventually built but production ceased in 1929.

The depression years were a lean time for speculation but Westland kept going by re-engineering the DH9 biplane with a new fuselage and a Bristol Jupiter radial engine. This aircraft, named the Wapiti, was a commercial success and eventually more than 500 were built and used by the RAF in the north-west frontier of India and in the deserts of Iraq. The company then proceeded with a private venture, which started life as a Wapiti mark V, with a supercharged Bristol Pegasus engine. It became the first aircraft to fly over the peak of Mount Everest in April 1933. This conversion received several contracts with the Air Ministry and more than 170 were built by 1936. However, in 1934 a sea change rocked the company, which had hitherto enjoyed a period of stability with Robert Bruce as managing director and Arthur Davenport as chief designer. This change was forced upon the company by the chairman, Sir Ernest Petter.

The strutted monoplane the Widgeon, designed by the chief designer, Arthur Davenport, in 1923.

Ernest Petter was a strong-willed, domineering father who had ambitions for his bright son, Teddy. However, Teddy saw more of his mother than of his father, since Ernest spent so much time in London, and from her he inherited a strong religious conviction and a set of firm ethical principles, which played a major part in his upbringing. He was reticent and scholarly, and looked it. His appearance was to influence his relationships throughout his business and family life.[3] He might have looked a scholar and introvert, but he turned out to be supremely confident in his own abilities, almost arrogant, and was to have many conflicts with government officials and captains of industry, who might have thought he was a pushover.

Teddy was sent away to prep school until he was 12, when his father decided he should go to Marlborough public school. According to Teddy's brother Gordon, his father probably picked this prestigious public school as a reaction to his own lack of higher education, and to ensure that his eldest son had the best possible preparation for taking over the family business.[3] Ernest also wanted Teddy to become proficient in such sports as rugby and cricket. Teddy's first house master was a double blue and expected all of his pupils to demonstrate a similar prowess. Although Teddy proved to be an outstanding scholar, he did not enjoy any of the sports, having

The young Teddy Petter, possibly taken at Marlborough College or at Cambridge.

neither the requisite physique nor inclination. He is said to have preferred to spend time on intellectual pursuits, and was especially interested in the great philosophers Kant and Nietsche.[3] For relaxation he read motor car magazines and took solitary walks in the Wiltshire countryside. His time at Marlborough did much to develop his aloof and withdrawn attitude as a defence against the many snubs he must have endured by not being one of the sporting set.

He left school for Gonville and Caius College at Cambridge, where he read engineering. In the final year of the Mechanical Tripos he was awarded a first-class honours and a gold medal in aerodynamics. He led a reclusive life at college, but had one friend, John McCowan, who had rooms on the same staircase. They shared a passion for motor cars and acquired an old air-cooled sports car, in which they travelled to the McCowan's farm in Yorkshire. It was at this farm that Teddy met an attractive French girl, Claude Marguerite Juliette Munier, the daughter of a Swiss official at the

League of Nations in Geneva. She was to become his beloved wife, although his engagement was not welcomed by his parents, who were prejudiced against 'foreigners' and compared her unfavourably with the daughter from a well-heeled English family with whom his brother Gordon was seeking an engagement at the time. Gordon is on record as attributing his brother's puritan strain to their mother and her Huguenot ancestors.

On graduating in 1929 Teddy agreed to his father's suggestion that he join Westland as a 'graduate apprentice'. He then served the statutory two years moving from workshop to workshop, and is believed to have insisted on no special favours as the chairman's son. The chief test pilot, Lois Paget, said 'I can't stand that priggish young man.' In fact in later years Teddy was to say: 'I looked on this as sheer drudgery at the time, but knew afterwards that without workshop knowledge I would never have become a designer.'[3] In the spring of 1932, Sir Ernest appointed Teddy, at the age of 23, as personal assistant to the managing director, Robert Bruce, who perhaps understandably did not welcome this appointment and ignored his presence in his office. Teddy consequently spent much of his time with John McCowan modifying an Austin 7 to win several sporting trophies. Despite this enthusiasm for sports cars, he never wished to learn to fly. The Westland chief test pilot, Harald Penrose, tried to get him interested but reported 'his sole attempt at piloting revealed a lack of the required sensitivity coupled with hopeless judgement of speed and distance, so his efforts at approaching for a landing were abysmal'.[4]

Whilst serving his time as personal assistant, Teddy set the date for his marriage to Claude: August 1933 in her home town, near Geneva. John McCowan was best man, and Sir Ernest and Lady Petter attended somewhat grudgingly. Teddy's two daughters, Camile and Françoise, were born whilst he was serving out his assistantship with Bruce, whose retirement was due in late 1934.

It seems that Bruce had an inkling that Sir Ernest would appoint Teddy as his successor, so in May 1934 he announced his resignation, to take effect within a month, whereupon Ernest announced that Teddy had been co-opted to the board and would be in charge of all design activities. Ernest later entitled Teddy to assume the title of technical director. This looked like a deliberate snub to the chief designer, Arthur Davenport, and Bruce's son-in-law, Stuart Keep. Bruce insisted that Keep, at least, should also be appointed to the board, and when this was refused tendered his resignation to take effect immediately. Teddy became technical director at the age of 26, with the more experienced Davenport (aged 43) reporting to him, whilst Keep was made works manager. For several years the Air Ministry left Westland off their list of potential bidders for military aircraft because Petter was not considered a sufficiently experienced or dependable designer.

In fact Teddy had learned a great deal from Bruce in three years, especially his means of keeping tight control on design development by spending time at the boards of all of his draughtsmen and at the desks of his aerodynamicists and stressmen. Petter was not alone in this practice apparently as others such as Sydney Camm at Hawker and R.J. Mitchell at Supermarine operated similarly. Petter adopted this practice throughout his career and consequently was never hit by unexpected surprises. He was able to make informed decisions on who deserved promotion and who did not. One of his first important decisions was to scrap the Pterodactyl, a grotesque aircraft, which was essentially a flying swept-back wing and needed no tailplane or fin.

This was not a good time to expect profitable contracts for military aircraft. Sir Ernest gave a lecture to the Royal Aeronautical Society in Yeovil in which he emphasised the need to remove barriers leading to delays and heavy costs in the development of new aircraft. He asked for encouragement similar to that enjoyed by manufacturers on the Continent and elsewhere. Teddy was not, apparently, helped by his aloof attitude towards the Air Ministry officials, who found him supercilious and overbearing. Consequently, as no military aircraft contract looked to be forthcoming, Petter persuaded his father to put up company capital to

From left to right: Arthur Davenport, Harald Penrose, test pilot, and Robert Bruce.

pursue a private venture, a high-wing three-engined six-seater to compete with the De Havilland Dragon/Rapide.

Unfortunately Ray Chadwick at Avros had already proceeded with such a design and had interested Imperial Airways and the Air Ministry. So, in his first year as technical director, Petter tried his hand at a wide range of aircraft designs, all without success. However, the Widgeon, designed by Davenport ten years earlier as a high-wing strutted monoplane, was going through several development cycles and it allowed Petter to start his career as a designer. In particular his new design, called the P7 (P for Petter?), exploited an adventurous Handley Page device for high lift, namely a wing leading edge slat which automatically deployed as the wing incidence was increased and the leading edge pressure intensified (see Appendix 1). The movement of these slats was also geared to the trailing edge flaps.

Westlands showed that this system was highly effective and mechanically reliable. This established Petter's reputation as a successful designer. In 1935 the army co-operation squadrons were looking for a replacement for their Hawker Audax, and the Air Ministry included Westlands on its list of bidders for a military aircraft designed to have a very low stalling speed. Petter's response was the P8, later to be named the Lysander. Hawkers had submitted their own replacement design and were awarded a contract for 178 Hectors, which was passed on to Westland as a subcontractor: a blessing for Westland and Petter in particular. Westland were now building the Hector and the Lysander. (It is tempting to speculate whether the name Lysander, to join Hector, was the choice of the scholar Teddy Petter.) The Lysander's time was still to come, and it would eventually replace the Hawker aircraft in the RAF squadrons.

2

THE WESTLAND YEARS: THE LYSANDER, 1935–1944

On 16 July 1935 came the first of several events that would nearly break up the Petter family. At this time, Teddy was 27 and had been technical director for less than one year. Sir Ernest Petter had become the Westland chairman, and his replacement as managing director was Peter Acland, who had been in charge of the London office. Acland went on to play a major role in supervising production with Teddy's help, thereby undercutting the general manager, Stuart Keep, who finally felt forced to resign. Westland needed more new workshops so Sir Ernest decided to look for expansion opportunities. He convened a shareholders' meeting to vote an increase in Westland's capital to £750,000 by creating more shares to provide funds for a merger with British Marine Aircraft, who had large workshops at Hamble, on Southampton Water. Teddy and Acland threatened to resign, so he decided not to proceed with the merger.

This conflict between Teddy and Sir Ernest would take a long time to heal. Teddy still had a scholarly, almost reserved appearance, a feature which emphasised his reputation for arrogance. He did, however, have a stubborn streak, as was to become evident in future conflicts with those who did not share his views. In the present case the conflict was with his father. The rift was to become even greater in 1938, when Sir Ernest negotiated a deal with John Brown shipbuilders to buy a controlling interest in Westland Aircraft. In the meantime Teddy was to take his P8 design as a successful bidder for the Air Ministry's new military specification, A39/74.

The specific requirement for 'The Army Co-operation Aircraft' seems a rather curious one today. Its roles included general liaison duties (such as the transport of executive officers), limited bombing, message pick-up (scooping bags on ground), reconnaissance and spotting enemy artillery. However, in the 1930s all nations agreed on the need for rapid and efficient communication between senior military staff, especially as an army was advancing over a wide front. Today communication is very sophisticated: think of wars undertaken under the auspices of the United Nations, when more than a dozen nationalities, speaking several languages, have to fight a common enemy and not each other. This was not the case in the 1930s. Semaphore was still a conversational tool (and still is in the NATO navies). The design of an aircraft in this 'army co-operation' role, foreseeing a European war in the offing, was undertaken by several nations: Germany, France and Poland. Four are worth mentioning.

The earliest design was the French Brequet 270 in 1928. It was a biplane produced for the Armée de l'Air and had a high drag, which resulted in a maximum speed of only 147mph. Its range was 620 miles, a value considered desirable. A total of 100 were delivered to the armed forces in 1930 and 1932.

The Henschel HS 126 first flew in 1936. It was contemporary to the Lysander and the designers recognised that, in addition to short take-off and landing, it should be able to defend itself in the war zone. It therefore had a pilot with a synchronised machine gun, and a rear gunner with a free machine gun. It had a respectable top speed of 221mph at 10,000ft, but a range of only 360 miles. It enabled Luftwaffe 'volunteers' to take part in the Spanish Civil War, and saw service in Greece, Croatia and, later, the Eastern

German Henschel HS 126.

Polish LWS-3 MEWA (seagull).

German Fieseler Fi 156 'Storch.'

Front. More than 800 were built. It was retired in 1942 from active service and replaced with the Fieseler Storch.

The Polish LWS-3 MEWA was designed and built by Lubelska Wytomia Samolotow, using French propellers. It had a range of 440 miles, a maximum speed of 224mph at 27,000ft and a very respectable stall speed of less than 60mph. It was equipped with 8mm machine guns fore and aft, and had an unusual mechanism for lowering the fin and rudder when the rear gunner fired. Only two were built before the German invasion, and in September 1939 they were scrapped.

A most unusual army co-operation plane, and ambulance, was the German Fieseler Fi 156 Storch. It had a fixed leading edge slat and a slotted flap, resulting in an effective increase in wing area and an extraordinarily high lift coefficient. This gave the later versions a superb take-off distance of 100ft and landing distance of 160ft for a machine that empty weighed only 1,900lbs. Because of the fixed slats, its cruising drag was high, resulting in a maximum speed of only 109mph and a range of 240 miles. Another contributor to the drag was its long legs, consisting of oil-and-spring shock absorbers with a travel of nearly half a metre. The wings could be folded back along the fuselage for easy transportation on land. It was clearly a machine capable of landing in rough terrain, and became famous for rescuing Mussolini from an Italian mountaintop in the war. It was used by Rommel in the North African deserts, and also by several English field marshalls, including Montgomery, when they were captured. The Luftwaffe accepted 2,900 of these light aircraft.

The most important part of the British Air Ministry's specification A39/74 was the need for a short take-off and landing on unprepared grass fields. Petter himself was initially doubtful of their capability for meeting this specification, and so he personally spent some time consulting pilots in the existing army co-operation squadrons. Most of the pilots agreed that a replacement for the Hawker Audax was badly needed, capable of landing in, and escaping from, small spaces without losing control. For this the pilot needed a clear forward and side view, and Petter made this a priority in his design.

Petter's approach was to turn the configuration of the Widgeon into a robust military aircraft that could be produced in large numbers by the current labour force at Westland. He had already reorganised the drawing offices, sacking 'dead wood', of whom he said to Harald Penrose, 'they'll soon get their pensions'.[4] This led to delays in supplying spare parts to the Indian Air Force with which Petter had arranged contracts. Indeed Petter was proud of his talent for both designing and marketing. (Years later he was able to persuade the Indian Air Force again to buy Gnat fighters when the UK government decided not to.) Because of such delays, the Air Ministry threatened to move Westland from the list of established contractors, unless the manager, Bill Gibson, was sacked. Petter did not contest this request: apparently he did not like Gibson.

He did not seek to recruit more draughtsmen since he believed that once the basic design had been chosen and evaluated (by him) then all basic choices had been made, and what remained were detailed design and equipment choices. He had already established the practice of going daily to all the boards of draughtsmen, and to the desks of aerodynamicists and stressmen. Glen Hobdey, who later became chief stressman at Yeovil and Warton, was to say that 'everyone held Teddy in great respect, whether managers, foremen, charge hands or fitters. No-one had a bad word for him.

His experience seemed to embrace all the many disciplines involved and he could discuss everything with knowledge and authority'.[3]

A very young Hobdey had been given an aileron that the drawing office had designed with three hinges to stress. This structure is redundant and not solvable by statics. Hobdey spent almost three weeks with log tables to perform a minimum potential energy solution; when Petter paid a visit, he took Hobdey's solutions and tore them up before depositing them in the waste bin. He said to Hobdey, 'Go back to the drawing office and tell them to design this aileron with two hinges. Your calculations will be much more simple and you will know exactly the loads that are imposed.' (He might have added that the aileron would not now jam if the wing suffered large distortions.)

The Lysander was unique amongst competitive aircraft designs in being designed from scratch to meet the Air Ministry specification, rather than being a modification of an existing design. The photographs show several of the new features. The crew consisted of a pilot in front, equipped with two 7.7mm machine guns, and an observer in the rear cockpit (open, with a sliding canopy) who managed a Lewis gun.[5] It was essential for the pilot to have good views forward, down and sideways. The high wing arrangement was consequently chosen. This ruled out a retractable undercarriage so a fixed configuration was chosen. The wheels then needed internal springs all encased in streamlined spats, but Petter was able to turn all this to their advantage in many ways.

The main wing had supporting struts anchored onto the undercarriage. This meant that the wing bending moment (see Appendix 2) was a

Lysander at the Smithsonian Museum, Dulles Airport, USA. This view clearly shows how the pilot had a clear and downward view. The leading edge slats are deployed along the entire wing. (Photograph Ad Meskens)

This shows one of the first Lysanders to have horizontal winglets added to the undercarriage spats for carrying small bombs.

maximum at this support and would decrease both outboard and inboard from this point. The bending stiffness is proportional to the cube of the section depth, so the wing thickness and chord could therefore taper down to the wing tip and also down to the fuselage centre line. Petter took advantage of this and not only tapered the thickness but swept back the leading edge, all to give the pilot a better view. The wing was mounted above the pilot's head anyway, so his view was exceptional.

In the photo the leading edge slats are deployed, as are the flaps to which they are linked. It has already been said how this automatic linked mechanism controlling the deployment of both controls was attractive, but potentially risky and prone to jamming. However, Petter had shown the system to be mechanically sound and reliable. This was impressive technology for a 1937 machine. Because the automatic actuation of these controls depends on the high suction at the leading edge (see Appendix 1) it is essentially triggered by a high incidence, although the precise incidence is adjustable in the linkage. This means that the tail should not be raised during take-off, which is the usual way of reducing drag and accelerating to take-off speed. The same is true for landing, but here the increased drag is beneficial. The landing speed was in fact only 52mph, compared with the top speed of 237mph. A speed range of 4:1 was remarkable at the time. With a payload of 5,000lb, the take-off run was only 103yds with the slats and flaps operating. With these controls closed, the run was more than double, at 226yds. The Lysander was therefore able to operate easily within the Air Ministry's airfield length requirements.

Petter had formed an alliance with W.C. Devereux, director of High Duty Alloys at Slough, who was able to supply a wide range of extruded sections. The unique undercarriage was a continuous square section bent into an inverted 'U', to which the internally sprung wheels were attached. This inverted U shape exploited its bending flexibility to cushion the loading if landing heavily. There was therefore no need for the usual oleo-pneumatic shock absorber struts. The concept of internally sprung wheels was conceived by a young designer from Cheltenham, George Dowty, who was developing a company called Dowty's that later became world famous (and still is). The large spats over the wheels also housed the forward-firing Browning machine guns. It is clear from the photograph that their firing

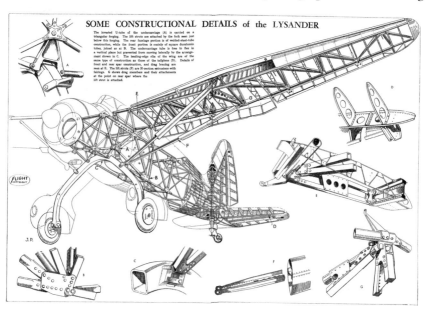

Cutaway of the Lysander showing the structure is essentially a series of frameworks with a myriad of special joints. (Courtesy of *Flight*)

line was outside the diameter of the propeller, so there was no need for an interrupter gear. A pilot may have only half a second to fire these machine guns at an approaching enemy aircraft, so the rate of firing is as high as possible. There was no room for ammunition in the streamlined spats, so ammunition to these guns was fed from the fuselage boxes down inside the square section undercarriage struts. Small horizontal and detachable winglets were added to the undercarriage struts, each carrying four bombs. An additional benefit of this unusual placing of armaments was their clear accessibility. 'Bombs up' and machine gun adjustments, plus replacing spent ammunition, could all be done by ground staff as everything was only 1 to 3ft above ground level.

Petter had not yet abandoned the familiar structural framework in favour of a stressed-skin construction. In the US, two companies had already pioneered the use of light alloy sheet to provide a smooth exterior as well as acting as the load-carrying structure. In the early 1930s the Boeing 247D and the Douglas DC1 used the stressed-skin concept for the fuselage, wing and all controls. All surfaces which had to take compression needed to have stiffeners added to prevent buckling. Petter was undoubtedly aware of these inventions, but reckoned the low operating speed of the Lysander did not merit a superior surface. He was probably also influenced by the need for accessibility and ease of maintenance if the primary structure was an open framework. Such frameworks are cheap to assemble by a relatively unskilled workforce, whether the framework joints are welded or bolted. It is also a damage-tolerant structure, being 90 per cent fresh air. (The ultimate example of a fuselage made as a geodetic framework was Barnes-Wallace's Wellington Bomber.) The front part of the Lysander fuselage used the type of construction with which Westland had much successful experience in their biplanes, that is a framework of Duralumin tubes of square section joined by bolts and gusset plates. The rear part of the fuselage used the new seamless steel tubes from Reynolds Tube and Co. Ltd, joined by welding as shown. This construction was quick and cheap to produce and all the tubes were lightly loaded and thin-walled. However, Petter was conscious of the drag for fabric-covered frames. The front fuselage was covered by detachable metal panels whereas the rear fuselage had thin wooden strip stringers to support the fabric cover as a smooth surface.

The wing structure was a compromise. It was known that the maximum suction pressures were at the leading edge (the net resultant lift acts through the quarter chord point) so the main structure was restricted to this region.

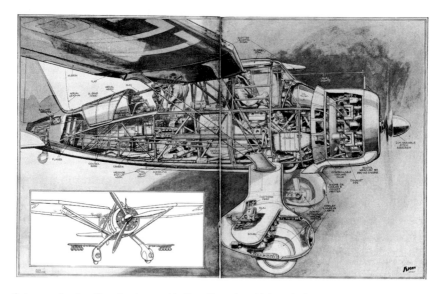

Cutaway showing the wing tapered both outboard and inboard of the strut, the sliding canopy over the rear gunner, and the Bristol mercury XII 9 cylinder engine cooled by propeller wash and exiting through controllable gills. (Courtesy of *Flight*)

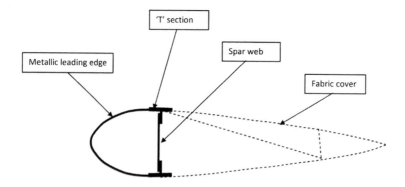

Schematic of the main wing structure showing how the load-carrying structure is confined primarily by the leading edge.

The front spar had two extruded 'T' section booms as beam flanges, top and bottom, joined by a vertical flat metal web. This web had regular stiffeners to prevent shear buckling (see Appendix 2). This beam was bolted to a curved light alloy leading edge, forming a closed tube, so providing a basic structure that had both adequate bending and torsional stiffness (see Appendix 2).

The rear portion of the wing behind the front spar was a light framework, sufficiently stiff to support the rear spar, flaps and ailerons, and covered by a fabric. It seems Petter was aware of the importance of correct flow over the leading edge, compared with the flow over the rear surfaces, but this is not as important as flow over the rear surfaces.

Although Petter was clearly an able mathematician, he did realise the limitations of theoretical aerodynamics and so he depended to a large degree on wind tunnel testing. This also had its limitations, particularly the ability to simulate accurately the effects of the slipstream from the propeller on the flow over the wing and tail surfaces. Often, anomalous and unrepeatable wind tunnel results were obtained for stability and control measurements under high-power conditions. There are reports that Petter was known to lose his temper with the wind tunnel staff on these occasions, but once consistent and repeatable results were obtained he became a dedicated believer, sometimes preferring test predictions to the findings of his own test pilots, who in those days could not always back up their opinions with data from on-board instrumentation.

The Lysander engine was originally a Bristol Mercury III, 890hp, 9-cylinder, radial-cooled engine driving a DH variable pitch propeller. The engine was prone to overheating, so cooling was introduced by controllable gills. This engine was later replaced by a higher-powered Perseus XII engine on an improved engine mounting to the fuselage.

The proposed design was approved by the Air Ministry in June 1935, with a contract for two prototypes. Within twelve months the first of them was ready for taxiing trials on 10 June, then moved to RAF Boscombe Down, and on 15 June was flown by Harald Penrose, who landed it back at Westland's airfield.

Initial test flights invariably reveal some problems. The first one was when Penrose said he ran out of elevator when only 10ft up because the tailplane was not adjustable. More seriously, he reported a measure of longitudinal instability when flying hands-free and with the engine throttled. On heading for his landing at Boscombe Down 'it was immediately clear there was insufficient tailplane adjustment for I was pulling back on the stick to maintain a slow enough speed when flaring out using little engine. The stick was already at its limit. I gave a burst of engine to increase the slipstream over the elevators and she settled gently, but the tail was still off the ground … The young Petter looked relieved at this first trial and certainly my report was favourable, but warned of the difficulties of tail trim.'[4]

An assembly line of Lysanders waiting for the next stage.

King George VI with Petter and Mensforth walking between two Whirlwind fuselages.

The wind tunnel tests, as so often, had failed to allow for the large downwash coming from the propeller, and this meant several redesigns of trimming gear to ensure a big negative tailplane angle for landing. This had its own dangers.

The tendency for stick-free longitudinal instability was eventually corrected with a larger tailplane, but the lack of control at landing remained. The danger was that it was impossible to hold the aircraft level when opening up the engine after an abortive landing due to the powerful effects of the slipstream over the tail. Even at half power it required all the pilot's strength to hold the control column fully forward whilst he was trimming with his other hand. The nose could zoom up into a stall, sometimes with fatal results, but Petter never really accepted that this was a dangerous feature of his design. He was determined to display the Lysander at the SBAC air show, and told Penrose not to tell Sir Ernest of the problems. All the test pilots proposed that an automatic retrimming device should be designed and installed, but Petter rejected this recommendation on the grounds that production was so urgent that no further time could be spent on development. The Air Ministry agreed, despite the protests of its own pilots, and awarded an initial contract for 144 Lysanders in 1936. The tailplane problem was eventually overcome by a variable incidence tailplane and a change in the handling technique for landing.

After solving the engine cooling problem, full production was continued. The Lysander became the main product of the Westland works from 1936 to 1938. Production peaked at eight per week, and a total of 1,368 were delivered to the RAF. This rate of production is suggested in the photograph of the assembly line, with more than two dozen aircraft in various stages of completion.

The first Lysanders entered service with No. 16 squadron in June 1938. They were the first British aircraft to be based in France at the beginning of the Second World War, and Lysanders of No. 4 squadron were the last of the British air component to see action during evacuation from Dunkirk. Pilots and crew were so pleased with the aircraft's handling capabilities that it soon became affectionately known as 'Lizzie'. Virtually all of the technical innovations that Petter introduced had worked. Operationally the slats and flaps gave the Lysander its extraordinary short field performance. On one occasion King George VI visited the Yeovil plant when a high wind was blowing over the airfield. Harald Penrose flew into the wind and was able to hover over one spot for several minutes.

In operations to provide close support to the British Army in France in 1940, the Lysander could perform its role of artillery spotting and bomb dropping very well, but due to its low speed and meagre armament, it was a sitting duck for enemy fighters such as the Messerschmitt Me 109. Petter felt that its single hand-trained Lewis gun could be replaced with a power-operated gun turret. But to place such a turret behind the pilot would cause a large destabilising shift in the centre of gravity that could not be corrected by the normal tailplane. A solution to this could be a complete redesign using a de Lanne wing instead of the tailplane. In fact Petter and Penrose went to see de Lanne in France and a prototype was made with the de Lanne secondary wing mounted beneath the fuselage just forward of the turret, and with full span elevators and twin endplate fins and rudders. The aircraft was tested by Penrose, who considered it handled well. None, however, were ordered.

The Lysander was a total failure in its primary role. The skies over France and Belgium in 1940 were simply too dangerous for a slow army co-operation aircraft. When the Germans attacked in May 1940, their armies were supported by swarms of Messerschmitt 109s. Allied fighters were overwhelmed. Of 174 Lysanders sent to France, eighty-eight were lost in aerial combat and thirty were destroyed on the ground. One hundred and twenty crewmen were lost. Only fifty-six aircraft survived to return to Britain. However, elsewhere, without the threat of the Luftwaffe, the Lysanders were more successful.

This Lysander was built with a shortened fuselage and a secondary de Lanne wing. The aim was to have a power-operated rear gun turret. (From *Aeroplane Monthly*)

The 208 squadron had received its Lysanders in April 1939. In Egypt the Italians entered the war in 1940, but did not pose a threat, and during the first successful phase of the war in the desert, the 208's Lysanders operated successfully in their intended army co-operation role. No. 6 squadron also used them in the desert, taking part in the advance west. However the 208 squadron moved to Greece in 1941 and came up against the Luftwaffe again. Of its nine aircraft, three were quickly destroyed, and the rest were withdrawn and replaced by Tomahawks.

The Lysanders also saw service in Burma and India with Nos 28 and 20 squadrons. No. 20 squadron was based in Assam, taking part in the siege of Imphal, and was the last front-line squadron to be equipped with Lysanders. They were retained by them until the summer of 1943.

The majority of Lysander squadrons were actually formed after the fall of France, performing vital air-sea rescue duties. Its low speed allowed it to drop dinghies and supplies to downed aircrew and it was also used for towing target tugs and for radar calibration. Of some 1,670 aircraft built, 961 were mark III aircraft, which first appeared in 1940.

A full listing of the many variants is worth summarising:

- **Lysander Mk II** Similar to the Mk I but powered by a 905hp Bristol Perseus XII radial-sleeve–valve-piston engine. This was considered to be a more robust engine than the Mercury, making it better suited to operating from the small fron-tline airfields that the Lysander was expected to use. This version was bought by L'Armée de l'Air (one aircraft), the Irish Air Corps (six) and the Turkish Air Force (thirty-six). Twenty were transferred from the RAF to serve with the Free French Air Force, and was the version in use in France in May 1940. One aircraft was supplied to Canada as a pattern for licensed production by the National Steel Car Corporation at Malton, Ontario. Seventy-five were built there.
- **Lysander Mk III** Similar to the Mk I, but powered by an 870hp Bristol Mercury XX radial-piston engine. Two hundred and fifty were built at Yeovil and seventeen at Westland's factory in Doncaster, Yorkshire. One was bought by the Royal Egyptian Air Force. One hundred and fifty were licence-built in Canada, powered by the Bristol Mercury XXX engine. This was the lowest-powered engine used by the Lysander, but the plane only suffered very slightly at low levels. The first Mk IIIs entered service in August 1940.

- **Lysander Mk IIIA** This was the final combat version of the Lysander. It had the Mercury XXX engine. Floor and side armour were added for the first time, and twin Browning guns replaced the single Lewis gun in the rear cockpit. A number of obsolete items were removed, such as a hook for picking up messages. A total of 347 were built by Westland; one was supplied to the Free French Air force, eight to Portugal and two to the USAAF.

- **Lysander TT Mk IIIA** The last 100 aircraft were custom built as TT Mk IIIAs, with their rear guns removed and a winch and pulley system installed. As many as 400 existing Lysanders were eventually converted in this way to act as target tugs or sea rescue machines as mentioned.

- **Lysander III (SD) and IIA (SD)** The SD designation was given to aircraft converted for work with the Special Operations Executive for clandestine operations. They were intended to carry agents or VIPs to and from enemy-occupied territory. Long-range 150 gallon fuel tanks were fitted beneath the fuselage, and gave the aircraft an endurance of eight hours. An access ladder (with luminous paint) to the aft cockpit was mounted on the port side of the fuselage for quick entry and exit at night. All under-surfaces were painted in a sooty non-reflecting paint. Guns were removed from this aft cockpit. These aircraft served with Nos 138 and 161 squadrons, and carried out at least 400 operations. The Lysander became mostly famous for this work with the SOE, and featured in several novels and films. This role therefore deserves a special account.

A Black Moon Lysander with an extra fuel tank, a ladder attached for rear cockpit access, and matt black under-surfaces.

Special Operations of the Lysander[3]

Almost immediately after the evacuation of the British Army at Dunkirk, a large intelligence organisation, known as the Special Operations Executive, was set up in London to co-operate with the underground movement in France. The Lysander was considered to be the ideal aircraft for clandestine operations with resistance movements on the European continent. Although agents and supplies could be dropped by parachute from other aircraft, only the Lysander could accomplish the return of agents and vital intelligence material to the United Kingdom. All it required was a long and level field for landing and take-off operations remote from occupation forces, a clear moonlit night and communications with the local resistance movement.

A special Operations Flight, No. 419, was formed in August 1940 with two Lysanders. Training began early in 1941 from a bomber airfield located on Newmarket racecourses. Then a few trial landings and take-offs were made in France, and finally the feasibility of the Lysander for these operations was proven in September of that year by taking in one agent and bringing back another. By then about twenty Lysander Mk IIIAs had been modified at a special plant at Ilchester, north of Yeovil, under a secret contract to Westland issued by the Ministry of Aircraft Production. Petter, of course, was informed of the purpose of these modifications but the secret was very tightly held inside the company. As mentioned, these aircraft were designated as Lysander Mk III (SD)s, standing for Special Duties. They also became known as Black or Moon Lysanders.

To maximise the Lysander's carrying capacity and range, all armament and armour were removed and most fittings from the rear cockpit. Normally not more than two passengers would be taken, but four could be accommodated in an emergency. Flight tests at Boscombe Down in 1941 showed the radius of action would be 450 miles at a cruising speed of 165mph. With two passengers, a 300yd field would be adequate. Training was necessary on the other side of the channel to provide agents who would select a suitable field, and lay out the three torches that constituted the flare path, which had to be closely in line with the wind direction.

The first operation was made in August 1941 to deliver one agent to the Chateauroux area in the middle of France, a distance of 270 miles, and bring out another, who was a former French liaison officer with the British Expeditionary Force, and who had been recruiting agents in France. His place was to be taken by a British ex-insurance broker who had friends in that part of France who might be enlisted for sabotage duties.

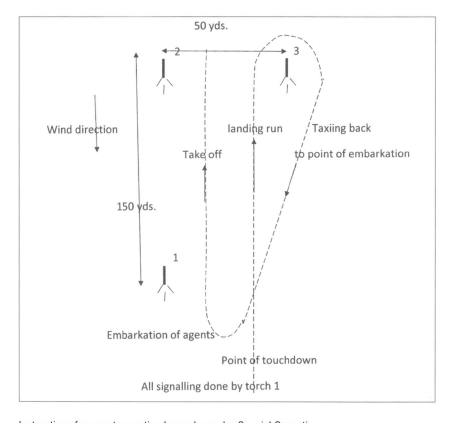

50 yds.

2 3

Wind direction   landing run   Taxiing back

Take off   to point of embarkation

150 yds.

1

Embarkation of agents

Point of touchdown

All signalling done by torch 1

Instructions for agents meeting Lysander under Special Operations.

This first operation nearly ended in disaster. The pilot arrived on time but the returning agent had been delayed by local police carrying out a routine check of all identification papers. On approaching the landing area on his bicycle, he found the Lysander already circling and realised that if fuel was running low, the pilot, with no signal from the ground, might decide to return to England. Desperately, he picked on another field he was passing, and with his helper they set out the L-shaped landing pattern with their torches. The pilot saw the pattern, landed and taxied to the other end. The agents quickly exchanged places and the pilot took off after only four minutes on the ground. The take-off from this new field was very rough, but worse still was the obstruction of a screen of trees at the end of the field. The pilot found what he believed to be a gap to fly through, but there were power and telephone cables that were not visible to him. There was a blinding flash as the power cables were severed, but fortuitously no serious damage occurred to the Lysander or the crew and it returned safely to England, trailing telephone wires all the way.

In January of the next year a mission to a field south of Tours led to a successful exchange of two agents in each direction, but sudden and massive cumulous cloud patterns were encountered and, as the fuel remaining was marginal, the pilot attempted to fly straight through severe turbulence. Icing on the wings forced the pilot to make an emergency landing in a rough field, at the end of which the Lysander ran into a ditch and turned up on its nose. The pilot and agents discovered that they were in fact still south of Tours, and quite close to where they had taken off. They re-contacted the local underground and found an alternative way back to England.

The experiences of similar sorties led to very few failures and a high success rate, despite the fact that crashed Lysander had made the German Intelligence aware of these Special Operations and their modus operandi. Sorties continued successfully through 1942, during which time fifteen agents were taken in and nineteen brought out. By 1943 the number of Lysanders on Special Operations had increased to nine and a total of thirty-eight sorties were made that year, delivering sixty agents and bringing out eighty-one. The only aircraft losses occurred in December, when there were two fatal crashes in thick fog in Southern England.

By the end of September 1944 nearly the whole of France had been liberated and no further covert activities were needed across the channel. Altogether the Black Lysanders had operated 400 sorties. Number 161 squadron alone took 293 people into France and retrieved 500. The Black Lysanders continued to operate in the Mediterranean areas up to VE day, with sorties in Italy, Yugoslavia and Greece; and behind the Japanese lines in the Far East until VJ day.

In summary, the Lysander was undoubtedly a technical success, exceeding all that the ministry specification asked for. Commercially it was also a success for Westland, who sold nearly 1,700 aircraft. Petter's philosophy of 'keep it simple' really had worked; that is, the aircraft were cheap to make and robust in service. This was confirmed by the large number of sales overseas, including more than two hundred that were built under licence near Toronto for the Royal Canadian Air Force. It was, however, vulnerable, with its low operating speed. This could be deemed an error in the ministry's specification, especially an underestimate of the performance of the Me 109s.

3

THE WESTLAND YEARS: THE WHIRLWIND AND THE WELKIN, 1938–1942

As mentioned previously, the rift between Sir Ernest Petter and his son came to a head in 1938 when Sir Ernest negotiated a deal with John Brown shipbuilders to buy a controlling interest in Westland Aircraft. It was thought that this deal was unfavourable to the Westland shareholders.[3] At the same time his brother Percy negotiated the sale of Petter Engines to Brush Electrical Engineering. The entire engine plant was shipped to Loughborough, thereby releasing extra factory space needed for the increased aircraft production. Sir Ernest had achieved both his financial and production aims and was content to step down, as he had suffered a number of fainting spells due to his low blood pressure. His doctor advised him to lead a quieter life.

Teddy saw the loss of family control as the loss of his birthright. He was furious with his father and even went to the extent of visiting his father's clubs in London to denigrate his character. He also reported adversely on his father's conduct to the Air Ministry. At this time the government was beginning to take aircraft production seriously as a national priority and it carried out an investigation into the ethics of the John Brown takeover. Although no wrongdoing was discovered, Sir Ernest's dream of having his knighthood converted to a baronetcy would never be realised. He retired to British Columbia, where, during the war years, his home would be a haven to Teddy's two older daughters. Teddy's brothers and sisters took sides in the dispute and the Petter family remained sadly divided for many years. Before Sir Ernest's death in 1954 the rift was healed to some extent and he finally left his money equally divided between his four children.

The instigator of the deal with John Brown was Eric Mensforth, son of Sir Holberry Mensforth, a director of John Brown. He had been working for a steel fabrication company that had been acquired by John Brown and Co. and his knowledge of Westland came as a result of his supervision of the construction of a large assembly shop for Westland. His message to John Brown was that Westland, with its Lysander work, would be a sound acquisition.[6] This reached Chairman Lord Aberconway and the John Brown Board, who then appointed their own people to the Westland Board. Sir Ernest resigned and was replaced as chairman by Aberconway. At the first board meeting Eric Mensforth was made joint managing director with Peter Acland. Petter remained as technical director. Naturally Teddy could never become a close friend of Eric Mensforth. He had lost his father's support and had to tolerate an appointee of John Brown looking over his shoulder on a day-to-day basis. Peter Acland also resented this and he resigned in 1939, leaving Mensforth as the sole managing director. The cast of characters for the war years was therefore:

Eric Mensforth	Managing Director
Teddy Petter	Technical Director
Arthur Davenport	Chief Designer
William Widgery	Experimental and Test Engineering
Harald Penrose	Chief Test Pilot

Nominally, at least, Teddy reported to Eric Mensforth, but they were to have a continuing clash of personalities for a variety of reasons. As Teddy would later admit to a reporter, 'I find the greatest difficulty in working under the authority of other people.'[3] Particularly so under Mensforth, who did not have any prior experience of aeronautical engineering. Both men came with preconceived ideas on how things should be done in the other's areas. Petter always felt that he was not getting enough support in the experimental development workshop area, which he strongly believed should be completely independent of the rest of the production plant and under the command of the technical director.

In fact he believed the technical director must take a wider-ranging view of the process of getting an aircraft developed and produced, and not just the technical and design functions. Thus he maintained a good current knowledge of how much each aircraft part cost, from which the cost per pound was also monitored. He also kept account of the manufacturing progress in both the experimental and the production shops. This provided the database for a stream of critical memos sent to Mensforth; justified or not, this was not conducive to the best of relationships between the two. Specifically, Petter always blamed Mensforth for not getting the Whirlwind into production early enough to participate in the Battle of Britain, although Rolls-Royce were actually the chief offenders.

Teddy always had an awkward relationship with Arthur Davenport, although he clearly recognised Davenport's vast experience and expected him to lead the design office after he (Petter) had taken all the initiatives. The two men were almost complete opposites. Petter had a Cambridge first and first-rate analytical and mathematical skills; Davenport had extensive design and manufacturing experience, based on starting every new aircraft as an incremental improvement on the last one, but lacked the analytical skills to take on and judge novel concepts, at a time when technology was making huge leaps forward.

Bill Widgery had no formal engineering training but seems to have been blessed with much common sense, and a native appreciation of basic engineering and physics. He was well suited to running the experimental tests and Petter accepted the laid-back attitude that was his hallmark most of the time. They got along well together except for occasional arguments about the wind tunnel operations.[3]

Petter's most valuable and influential colleague was the chief Test Pilot, Harald Penrose, and they worked together for the entirety of Petter's time at Westland. If a prototype behaved in an unexpected fashion, bordering on the dangerous, Petter said, 'You test pilots have to accept a few risks you know.'[4] He gave the impression that he often thought that Penrose, and his other pilots, were being too critical about the flying qualities of the aircraft, and consequently its acceptance by the RAF and the ministry was being slowed down. In truth Penrose and his fellow pilots were correctly anticipating what would be the opinions of the service pilots were they to test the aircraft at the relevant stage of development.

These then were the team and relationships that would exist through the design, development and production of the Whirlwind and Welkin, as well as through the wartime manufacture and servicing of many subcontracted aircraft.

In 1936, at the time when the Lysander was going into full scale production, the Air Ministry was becoming concerned at the increasing speed of British and German fighter aircraft. The time for a pilot to line up an approaching opponent, aim and fire his armament was coming down to one or two seconds. In such a short time the damage which could be caused by two machine guns would likely be insufficient to ensure a kill. Any effort to improve the efficiency of machine guns would hardly solve the problem. The standard Vickers 303 was already unreliable and prone to jamming. The Browning guns, which were a development of the American Colt design, were fitted to the Hurricanes and Spitfires in 1937 and 1938 respectively. The ministry wished to see more use of high-velocity cannon shells, either ball or high explosives, even one of which, exploding as a near miss, could penetrate armour plating. The Coventry Ordnance Works had developed the C.O.W. gun fighter, carrying a heavy 37mm cannon. However, Aero Engines of Bristol had obtained a licence to build a much lighter 20mm cannon of Hispano-Suiza design, and undertook a contract to supply for the Air Ministry. The rate of fire would be 650 per minute with a muzzle velocity of 2,900ft/sec. It was felt that a larger fighter than the Hurricane and Spitfire would be needed to control the recoil and aiming of these cannons. The ministry therefore issued the F35/37 specification for a single-seat, twin-engine day/night fighter to be armed with four such weapons. In two seconds this firepower could deliver 50lb of shells at an impact velocity of over 2,000ft/sec, which should be adequate for a certain kill on even heavily armoured aircraft. The specification for such a fighter called for:

- a speed greater than contemporary bombers, by at least 45mph at 15,000ft
- a maximum speed of at least 330mph at not less than 15,000ft
- sufficient numbers of forward-firing 20 or 23mm cannons in a short space of time
- maximum altitude not less than 30,000ft
- take-off over 50ft barrier in 600yds
- ditto landing in 600yds
- variable pitch propellers
- retractable undercarriage and tail wheel
- enclosed and heated cockpit
- night flying equipment.

Clearly this specification was a milestone in fighter development history, for not only was it the first formal British requirement for a four-cannon fighter, but it was among the front runners in the world to specify such a heavy armament.

It was probable that the Air Staff and Ministry were aware of the industrial competence in Europe and the US, in addition to the British contenders. Indeed as early as May 1935, under a Reich Aviation Ministry contract, Messerschmitt completed the first prototypes of the Bf 109 using Rolls-Royce engines, since the German Jumbo engines were not yet available. This aircraft became the most produced fighter in history with more than 34,000 being made. In February 1936 the director of technical development (DTD) proposed that the following firms should be invited to tender for a contract.[7]

- Armstrong: Design capacity very limited but clever at armament work.
- Fairey: Have already experience of 20mm cannon with their Fantome fighter. Very little design capacity now.
- Vickers: Design capacity limited at present, but good armament experience. A damage-tolerant geodetic construction may be possible.
- Westland: Have the design capacity and some experience, having built C.O.W gun fighters.

Eight established firms had been invited to tender to the specification, but only five had replied. An analysis of these tenders gave the following delivery dates:

- Boulton Paul (single engine), 15 months
- Bristol Type 153, 18 months
- Hawker (single engine), 20 months
- Supermarine (single engine), 20 months; (twin engine), 27 months
- Westland P9, 18–24 months.

The general opinion was that Mitchell's design, for a Supermarine twin-engine high-speed fighter, was a front runner, but the delivery time was excessive and their drawing office was already overloaded. Before discussing Westland's proposal (the P9) it is worth looking at the competition, bearing in mind the primary needs: a stable platform for four cannon, a short time to climb to altitude and a superior speed there, take-off and landing in less than 600yds.

Fairey Fantome.

Gloster F9/37.

Messerschmitt Bf 109E.

- **The Fairey Fantome** This was a biplane, first flight in June 1935, with one 20mm Oerlikon cannon. Maximum speed was only 270mph.
- **Bristol Type 153A** Four 20mm cannon; speed at 15,000ft, 360mph; rate of climb, 3,550fpm. A later model equipped with Fowler flaps could land on a carrier at 60mph.
- **Gloster F9/37** Two 20mm cannons; Speed at 15,000ft, 360mph; rate of climb, 2,400fpm.
- **Messerschmitt Bf 109E** One 20mm cannon; speed at 15,000ft, 338mph; at 17,500ft, 384mph. Service ceiling, 35,200ft; rate of climb at 30,000ft, 740fpm; stall speed with flaps and undercarriage down, 62mph.

Lockheed P-38 Lightning.

Westland P9.

- **Lockheed P-38 Lightning** This aircraft was not a contemporary American machine; in fact it was designed to a US Army specification in February 1937, and did not fly until January 1939. Its performance figures do show that the British specification was ambitious. Thus, maximum speed at 28,000ft, 443mph; rate of climb, 4,750fpm; a single 20mm cannon.
- **Westland P9** Max speed at 15,800ft, 354mph; service ceiling, 30,300ft; rate of climb at 15,000ft, 2,310fpm; four 20mm cannon; stall speed with flaps and undercarriage down, 85mph. These figures show a speed of 354mph, equal to the Spitfire I (and better at sea level), and rate of climb the same. What is not shown is the drop off of max speed and rate of climb at 30,000ft (145fpm compared to Spitfire's 1,020fpm), but more of this later. The stalling speed for landing also looks high.

At the Tender Design Conference the final order of merit was Supermarine, Westland and Bristol, with the DTD preferring the Westland submission as it was more technically advanced than the other tenders. The conference found the following points in favour of the Westland design:[7]

- The firm was not overburdened with design work.
- The Air Ministry wished to reward Westland for their successful Lysander design.
- Another twin-engined design was desired to offer competitive stimulus to the Supermarine team.

At the 1936 Design Conference the following prices were submitted: Boulton Paul £20,500; Supermarine £22,000; Westland £27,500 for the first aircraft and £18,000 for the second.

In February 1937 a contract was placed with Westland for two prototypes, to be delivered between August 1938 and February 1939. Possibly in recognition of the very advanced (possibly risky) design features in the Whirlwind, the Air Staff required the two prototypes to complete brief handling trials before a production order would be granted.

When Westland received their copy of the F37/35 specification it was found to include a number of alterations to the original issue, some of which were: armour to be 9mm thick; the aircraft fully loaded to withstand an impact with the ground of 12ft/sec; and the impact load on the undercarriage not to exceed three times the weight of the complete fully loaded aircraft. Petter was still a young man, aged only 29 years, and was probably viewed with caution by the ministry; how then did he succeed in designing such an advanced machine? The answer is that he viewed the challenge as needing a superior low-drag aircraft, built with a low-weight stressed-skin structure.

The P9 Project (Whirlwind)

Given that there were ten specified requirements in F35/37, Petter chose to take two as the most important. Firstly to mount four 20mm cannon on a stable platform able to take the high recoil whilst the pilot is taking aim for about two seconds, and to have a large number of high-explosive shells and be invulnerable to even machine gun fire from an approaching opponent; secondly to have a high rate of climb and a high speed at altitude.

The first requirement was relatively easy to satisfy by mounting the cannon immediately in front of the pilot and having three armour plates to protect the drum magazines and the pilot.

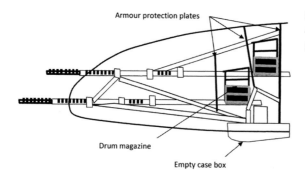

Armour protection plates

Nose fairing containing the 4 x20mm. cannons and the protection plates.

Drum magazine

Empty case box

The cannons would be held in a stiff framework, giving easy access for servicing the armaments. The assembly would be mounted directly in front of, and below, the fighter pilot.

The second requirement would be achieved, in addition to having adequate engines, by reducing the drag in as many ways as possible. Petter had never before taken minimising drag as a priority, and he resorted to many innovations, some radical.[3,7] The two engines were a given size and volume, driving three variable-pitch propellers, so they had to be enclosed in decent streamlined nacelles. The nacelle sizes were reduced to a minimum, and both Davenport and Petter felt that it was possible to make a great improvement in engine installation by placing the oil and coolant radiators in the wings. Rolls-Royce was sceptical of this proposal and their opposition was quite strong. Nevertheless, this strategy was later copied on the de Havilland Mosquito and the Hawker Tempest.

The fuselage had to have a low drag so his design had a smaller frontal area than the nacelles. The viscous drag (see Appendix 1) had to be small, so the entire aircraft was to be a thin-walled stressed-skin construction with smooth external surfaces. This was a huge advance on the Lysander, which was still being produced in large numbers. The outer stressed skin would be supported by a substructure of spars, ribs, frames and bulkheads and could easily be viewed today as a contemporary modern design. Only the controls had a fabric covering.

The first design for the P9 had a conventional tailplane and elevator with two fins and rudders mounted at the tips. This was later changed as wind tunnel tests showed that the wake behind the deployed flaps caused turbulent flow over the tailplane which would make controlled landings difficult. The tailplane was then moved to a position two-thirds of the way up a single central fin, well clear of this turbulence.

The new positions for the light alloy radiators were behind the leading edge of the main wing centre section, the inflow of air optimised by shaping of the two entry lips, but not too drastic as to interfere with the flow over the wing.

The engine exhausts were routed through the nacelle and then through the wing fuel tanks to emerge as ducts along the wing's trailing edge, and so provide a small degree of thrust augmentation, as compared with conventional stub exhausts. The test pilot, Penrose, was naturally horrified at this proposal, feeling that a bullet-damaged fuel tank could lead to a conflagration. This danger was realised in June 1939 when, during a test flight at Boscombe Down, the Whirlwind suffered a failure of the port aileron control at 8,000ft. The aircraft went into a viscous roll as this aileron went up to full lock. Penrose held on to the starboard control and got the aircraft back on to an even keel, and landed safely in a cross wind. A portion of the exhaust ducting was found to have broken away, which allowed heat flow around the aileron light-alloy push-pull rod, causing it to fracture under load. The ministry insisted on a return to the external type of stub exhaust system.

The main wings had an aspect ratio of eight, which was not bad for a fighter, and would have a tolerable induced drag. The main wing controls on this aircraft exploited some of the devices proven on the Lysander. The leading edge slats were of the Handley Page type and were fitted in two portions, the root slats being fitted above the radiator air ducts. The outer slats extended the full length of the outer wing leading edge to the tip of the wing. To ensure correct synchronous movement of each pair of slats, an

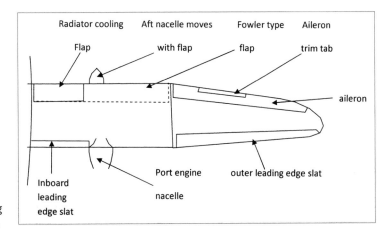

Port wing controls.

interconnection by means of a Duralumin torque tube was made behind the front spar. Fowler flaps (see Appendix 1) were fitted along the whole length of the trailing edge between the fuselage and the nacelle. Movement of these flaps automatically controlled the radiator shutters, and also opened the root leading edge slats when the flaps were lowered to more than a quarter of their maximum travel. The rear portion of the engine nacelles was attached to move with the flaps when activated hydraulically (a modern concept in civil aircraft for after-bodies covering the flap actuators). The aileron trim tab had a servo action and was also adjustable on the ground. This host of innovations might have alerted the ministry to possible development problems.

Petter's aim for minimum drag was compromised a little in choosing the basic configuration. Ideally the engine and nacelles would be cantilevered in front of the wing and placed symmetrically above and below the leading edge. Similarly the main wing would be mounted midway up the fuselage. Instead Petter chose to mount the wing at the bottom of the fuselage as in a normal single-engined fighter: thus the wing structure (and the bending moment) is carried continuously through the fuselage. The engines were cantilevered forward but below the wing box structure, thereby leaving the wing structure with a continuous form throughout. The undercarriage was also retractable into the nacelle behind the rear spar.

The airframe was designed to support a stressed-skin surface, which would carry as much of the required load as possible, since both the bending and torsional stiffnesses benefit from having material based as far as possible from the centre of the section. The leap in structural design from the frameworks of the Lysander to the stressed skin of the Whirlwind was truly monumental. The new structural design features, their sizes, and the material properties, had to be assessed before manufacturing the prototype. Petter had access to the American data sheets for stressing components, but the totally new configurations posed uncertainties: for example, how to transfer the engine loads to the wing. Traditionally stressmen guessed a 'load path' through a structure, that is to design a rib to take the engine shear, bending moment and torque and then to react these loads as smooth stresses connecting the rib periphery to the wing skins. The 'load path', which was assumed, may of course have been inaccurate: ignoring material away from the load path. But this error is on the safe side. If this ignored material took unexpected loads then it might fail and shed load to the assumed load path. Designers had well-known formulae for designing beams in bending and tubes in torsion (see Appendix 2) and this led Petter's team to design two structures: one, like spar booms, to take bending; and two, thin-walled tubes to take torsion. The 100 per cent stressed skin was yet to come, but at least a smooth skin surface with countersunk rivets was achieved.

Petter understood that a simple unsupported skin may need stiffening for compression loading, and he took full advantage of his new alliance with W.C. Devereux of High Duty Alloys, who would supply him with light alloy extrusions of almost any sectional shape. He compiled a large database of the strength and stiffness of castings and extrusions and he took these when he left Westland. In fact he had much expanded versions when he eventually moved to Folland, at which point the ministry gave him a contract to publish a collection.

The fuselage was divided into two sections with a transport joint at the rear of the cockpit, where the frame had a bulletproof steel armour plate.[7] The light aluminium fuselage skin was stiffened by vertical frames and longitudinal extrusions. A strong frame, set at an angle, provided support for the instrument panel and gun sights. The front nose fairing was detachable for servicing the cannons and drum magazines. The fairing itself consisted of longitudinal strips of light alloy sheet, joined at their edges at flanges of T-section extrusions. Duralumin frames maintained the shape of the fairing, and dowels provided location for connecting the fairing frame to the main fuselage. The rear fuselage was built up of magnesium alloy strips joined together and riveted to longitudinal T-section stiffeners. These strips were given a small curvature by applying a light pressure into wooden moulds at 300°C. This was a rather complicated way of forming an oval tapered cylinder but industry did not yet have the large metal stretch-formers for producing tapered cylinders and panels. The lipped vertical frames in Duralumin were riveted to the skin and had cut-outs to allow the longitudinal stiffeners to pass through. The tapering fuselage extended to an angled frame riveted to the fin main spar, and the fin substructure supported the rudder in the same fashion as the main wing ailerons.

The main wing was constructed in three parts. The centre wing passes cleanly through the fuselage, to which it is joined at the intersection by a strong bulkhead behind the nose section containing the cannon-supported framework. Two auxiliary spars were fitted in this centre section, the rear spar and ribs supporting the flap hinges. This wing centre section contained the fuel and oil coolant tanks, and also the oil and coolant radiators. These radiators were mounted on both sides of the front spar, whose web contained lightening holes to accept the air flow from the wing leading edge intakes.

The two 67-gallon fuel tanks were in this section, port and starboard, to supply fuel to a collector tank, but each system was separate and no cross feed cock was provided. This system was criticised by the ministry and the pilots, who could not use the fuel from an unserviceable engine to get back home. The two outer wing sections were tapered but the main spar was continuous with the centre section spar. It had aluminium extrusions as the two booms, tapering towards the tip and joined by a Duralumin web (like the Lysander). There were ten supporting ribs, the third and eighth of which carried the aileron hinges.

The tail unit was also a stressed-skin construction, the tailplane main spar joining the fin spar at a casting. The lower end of the fin carried the tail wheel unit. Both rudder and elevator were mass-balanced and had spring tabs to lighten the forces needed by the pilot (see Appendix 1).

The main undercarriages were stored in the rear of the nacelles. Both the main and tail undercarriages were fully retractable, and raised and lowered by a Dowty hydraulic system.

The contract was signed in February 1937, and wind tunnel tests were conducted at the Royal Aircraft Establishment later that year. They confirmed the low drag of this design and predicted a top speed of 350mph at 15,000ft. This was 35mph faster than the specification.

By early January in 1938, a mock-up of the Peregrine engine and its installation had been assembled and work on the fuselage and wings for the prototype was well under way. However, no engine had yet been delivered, and in March Petter wrote to the Air Ministry to express concern that engine availability would delay the first flight and therefore delivery by February 1939, as promised, would not be possible. He received a reply to the effect that if he could not fly by December 1938, the programme would be cancelled. Rolls-Royce then undertook to supply in five weeks and by May engines were delivered for both prototypes. Cracks appeared in the fuel tanks, but this was rectified quickly. The first prototype was ready for inspection on 27 September. The ministry found a large number of deficiencies (including a need for a bulletproof windscreen), which Petter thought to be trivial items which would be corrected before production started. The aircraft was ready for first flight in October and taxiing trials were completed at Yeovil. It was then disassembled and transported to Boscombe Down, which had much longer runways. Penrose made the first flight on 11 October and found the aircraft very manoeuvrable with excellent low-level performance. It appeared that counter rotation of the propellers was not needed. However, as tests proceeded it was found that the top speed and rate of climb fell off rapidly at high altitude. This was due to the engines' superchargers not meeting their promised performance at altitude. This deficiency was never corrected because Rolls-Royce felt the Peregrine was becoming obsolete, and wished to satisfy the overwhelming demand for Merlin production for the Hurricane and Spitfire.

There was also a problem for landing, when the leading edge slats were supposed to move automatically and actuate the flaps, as in the Lysander. This system was not operating reliably, so the decision was taken to lock the slats and operate the Fowler flaps directly. This increased the stalling speed and led to an increased landing run. This would later compel the ministry to veto night flights and restrict the number of airfields for landing. It could all have been corrected by redesigning the slats and flap links, but more urgent problems had appeared. Both engines overheated and Penrose was forced to make an emergency landing at an airfield near Dorchester. More seriously, he found handling difficult in dives and there was juddering in tight turns. This was found to be due to interference and compressibility effects at the junction of the fin and tailplane. After extensive wind tunnel tests, and a few more trials, this was solved by introducing an 'acorn' type fairing, which became standard on all production aircraft. With this modification Penrose achieved 410mph in a dive, outperforming the Spitfire.

All these problems were adding to the delays in securing a production order. Petter felt that he did not have full control over the experimental

First prototype of the Whirlwind all in dark grey. Notice that the fuselage is more slender than the engine nacelles.

'Acorn' type fairing added to junction of tail plane and fin.

fabrication shop, which he thought should come under his management. A sense of the frustration at this time can be seen in a quotation from Penrose's autobiography:

> I flew the Whirlwind to Yeovil (from Boscombe Down) and the long process of modification and test began, with alterations taking anything up to a fortnight between flights, and delays also occurring because the engines were prototypes. There was a partial engine failure, continual trouble with steaming and high oil temperatures; the rudder fabric collapsed, the Fowler flap would not operate correctly, the root slat control failed, so did the wheel brakes and the Exactor hydraulic engine controls gave constant trouble due to air in the system. By the beginning of December only six hours' flying time had been achieved, and in the next five, comprising nine flights, no great progress was made because the problem of buffeting had not been solved and there was tentative indication of trouble in attempting dives.[4]

The Air Ministry wrote to Petter, 'I feel that the difficulties envisaged by you at Yeovil would not have been so great had there been that willingness to co-operate with the ministry which we normally experience with aircraft firms.' It must be recalled that in 1939 both sides were under great pressure to re-arm the RAF for the major conflict in the air with Germany that was now appearing inevitable. Although the ministry was clearly concerned, they asked that the prototype be sent immediately to the RAE for evaluation before a production contract could be issued. Although their pilots confirmed the list of flight problems, they must have believed that these could be fixed, because a production contract for 200 followed soon after the aircraft was returned to Yeovil. John Fearn, the works' superintendent, promised the first operational delivery by September 1939, a lead time of just nine months. Presumably Mensforth approved this promise, but it seems likely that Petter did not, since he would have realised that there were still several major technical problems to fix. These required the first prototype to be laid up for two months for modifications, including a new rudder. By March the second prototype joined in the testing, and on one flight the canopy blew off in a high-speed dive, narrowly missing Penrose's head. The aileron control failure mentioned earlier needed the conversion of internal exhaust ducts to external ejector-type manifolds.

The time and materials needed to manufacture one Whirlwind were equal to that for two Spitfires, but if the performance merited it then a single Whirlwind won because it needed just one pilot. In fact the Whirlwind outperformed both the Spitfire and the Me 109 in both speed and rate of climb up to 20,000ft, after which the engine superchargers let it down. It was also clear that the Whirlwind's stall speed of 85mph, compared to the Spitfire's 64mph, meant it needed longer runways.

In the promised nine months the increased workload impacted on the two prototypes. Penrose described this phase thus:

Amid a plethora of production testing, the trials and modifications of both aircraft continued. Alterations of servo-tab gear ratios, changes to the sealed leading edge of ailerons, adjustments of slat venting to prevent a jerk on opening, insertion of a fin-shielded horn balance on the rudder. Because of these and the engine defects, only twenty hours' flying had been accomplished in five months during which 250 major and minor alterations and adjustments had been made. There was also continuing trouble with cooling, and maintenance of the electrical and hydraulic systems was proving difficult because of inaccessibility. To discount development problems I flew the second Whirlwind in May to a publicity demonstration for MPs, and the press reported: 'The fastest time of the day was not put up by a Spitfire but by a secret twin-engined machine which streaked over from the West.'[4]

In fact, details of the Whirlwind were not permitted to be published by *Flight* and *Aeroplane* until 1942.

The first production aircraft trials at Boscombe Down took three weeks in November–December, during which time the aircraft was dived from 14,000ft at 410mph, confirming that the 'acorn' fairing had cured the juddering problems.

It was at this time that the relationship between Petter and his test pilot deteriorated: Teddy accused Penrose of being too finicky in his standards for prototype acceptance. One of the items in contention was the difficulty in getting the tail down at landing. This was the same problem that was never resolved for the Lysander. As it happened, Penrose had to take time off for a hand injury, and Petter brought in another test pilot, who was more optimistic in his reports, so much so that Petter used this to convince the ministry to give Westland a contract for an additional 200 aircraft.

The first production aircraft was delivered to the RAF and flew on 22 May 1940, two years and three months after it was contracted. Meanwhile Rolls-Royce had to review its commitments due to a shortage of machinery and skilled staff. The Peregrine production rate was increased but was not despatched until February 1940, and finally stopped after 290 had been produced in December 1940. From this point the Whirlwind was doomed. The German bombers were increasing their operational altitude to 18,000ft, at which height the Peregrine superchargers were losing efficiency.

The production aircraft were delivered to No. 25 squadron RAF in June. Thus five years had elapsed between the initial issue of specification F37/35 and the delivery of the first aircraft to the RAF. By October the Battle of Britain was over and Lord Beaverbrook reduced the number of Whirlwinds to the 114 that could be made with the materials in hand.

Petter sent a letter to Fighter Command:

The Whirlwind is probably the most radical new aeroplane which has ever gone into service. New ideas I am afraid, even with the greatest care, always mean a certain amount of teething trouble additional to that which any new aeroplane meets on getting into Service. I really do not think these troubles have been any worse than they were on, say, the Spitfire. The Whirlwind has been moved to no less than four different stations in its brief life, the last two of which have been more than 400 miles from here.

Air Marshal Sholto Douglas responded quickly; he moved 263 squadron to Exeter and made the Whirlwinds operational.

The delays causing the aircraft to miss out in the Battle of Britain would always remain one of Teddy Petter's biggest regrets.

Whirlwind Operations

Air Chief Marshal Dowding had said the pilots liked the Whirlwind and pointed out it was the only aircraft capable of attacking tanks in case of an invasion. Sholto Douglas, who took over as chief of Fighter Command and moved the 263 squadron to Exeter, sent a letter to Petter:

> The squadron has been formed for four months but still has only twelve Whirlwinds and is therefore not yet operational. The delivery rate of one aircraft per week was much too slow and the squadron strength must be brought up to sixteen right away. Westlands were concentrating on producing Lysanders which nobody wanted instead of Whirlwinds which were needed badly.[6]

Although Exeter was quite close to Yeovil, it was quite unfit to support a high-performance aircraft. Nevertheless, Westland made a rapid response and by December about thirty sorties were flown from Exeter, although three aircraft and their pilots were lost. The proficiency and confidence of the pilots improved during the first three months of 1941, and one Arado 196 and two Junkers 88 were shot down over the English Channel. However, two Whirlwinds were lost. Due to the poor state of the airfield at Exeter, some aircraft were moved to St Eval, on the north Cornish coast. A Heinkel 111 dropped four bombs on this new site and damaged seven Whirlwinds. All were repairable but again operational status was set back. In April 1941 the squadron was moved to Filton, Bristol, but very little action was seen, and four aircraft

One of the first production aircraft, before the four cannon had been fitted.

were lost in accidents. Then in August the turning point in Whirlwind operations was reached when its mission was switched from interception engagements to attacks on ground and sea targets.

Attacks were made on airfields and parked aircraft near the French coast, and on the way there and back on 'E' boats and tankers in the channel. Many Junkers 87s and Me 109s were destroyed, and the German response to these attacks was to engage in dogfights. Four Whirlwinds were engaged at low altitude by about twenty Me 109s; three enemy aircraft were shot down and one damaged, with only slight damage to a Whirlwind engine. This was the first major success for 263 squadron, and they were then assigned to escort Blenheims on bombing attacks in France and Germany. In September a second Whirlwind squadron, No. 137, was formed in East Anglia. By the end of 1941 Whirlwinds had destroyed sixty-one enemy aircraft, with the loss of only seven Whirlwinds. Through 1942 and early 1943 attacks on airfields and shipping continued, and later night sorties on ground targets and convoy protection patrols were begun.

The Whirlwind was declared obsolete in January 1944 and squadrons re-equipped with Typhoons. So, as in the Battle of Britain, the Whirlwind played no part in the Allied invasion of Europe.

In summary the main problem was the poor performance over 20,000ft and the very long delays in reaching production status. This was due almost entirely to Rolls-Royce's poor support for the Peregrine engine and its supercharger. The design flaws had been described by the ministry as the high landing speed and the need for long runways, together with its unsuitability as a night fighter. It is of interest then to quote Victor Bingham, who managed to correspond with or interview many of the pilots who actually flew the Whirlwind.[7]

Air Commodore G.J.C. Paul said:

All my flying was as a guest of 263 squadron at Warmwell. The pilot's outlook taildown on the final approach was certainly much better than any of the Griffon Spitfires and Warmwell was not a big aerodrome. Was it a pilot's aircraft? Yes! I enjoyed every minute of it. The cockpit was comfortable, the view superb in all directions, controls nicely placed and came easily to hand, and the Peregrines as smooth as little dynamos.

Another Group Captain, J.B. Wray, DFC, ex-Commanding Officer of 137 squadron, said:

The Whirlwind was one of the most pleasant, delightful and safest aircraft I have ever flown. It had many features which were way ahead of their time in fighter development, the four cannons and the big tear-drop hood being but two examples. It was highly manoeuvrable and the two engines were a godsend bearing in mind one spent a lot of time being shot at by flak over the sea and enemy territory. Many a pilot got home on one engine. It had no nasty habits like many other aircraft I have known, and I flew over 100 different types including all the Marks of Spitfires. I never knew a pilot who operated on the Whirlwind to have anything but deep affection for the aircraft, and they would never hear a word said against it.

Another ex-Commanding Officer, Wing Commander H.St.-J. Coghlan, said:

The Whirlwinds had a reputation for being tricky to fly, especially at night. I found, and every one found, they were a beautiful aircraft day and night and there were no vices attached to them at all. Its flying characteristics were absolutely marvellous, and its reputation completely erroneous.

After the war, nine Whirlwinds were bought by a South American British group known as the 'Bellows fellowship' and one was bought by Mr and Mrs Ellis of Fiji, carrying the name *Comrades in Arms*.

The Welkin

One day in 1940, an air-raid alarm was given at Westland,[3] and the probable cause was a twin-engined aircraft flying at a great height in the direction of Bristol. Somewhat lower, what appeared to be a single Spitfire could be seen. Small puffs of smoke were followed by the sound of gunfire. The pursuing fighter then fell away and lost height, and by the time it was on course again, its quarry had disappeared from view. It was later said that the bomber was a Junkers Ju 86 on its way to bomb Bristol. The Spitfire had been so near its ceiling that the gun recoil had been enough to lower the speed and to precipitate a stall, and the recovery action cost it the loss of height and speed.

The maximum speed of the Junkers was 260mph at 30,000ft and the rate of climb 900fpm, with a service ceiling of 42,700ft. The Air Ministry was aware that the Germans had been developing, amongst others, the Junkers 86 high-altitude bomber, which with Juno diesel engines and a pressurised

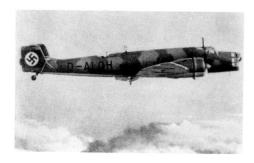

Junkers Ju 86. (Courtesy of Bundesarchiv)   The GAL 41.

Vickers 432.   Hawker Typhoon.

cabin had the capability to penetrate the RAF defences at altitude above the ceiling of their fighters. It therefore issued specification F4/40 to cover the design and development of a high altitude interceptor that could be deployed against aircraft of the Ju 86 category.

The concept of providing pilots with a pressurised cabin for extended stratospherical flight had started with much experimental work in Britain in the 1930s. From 1938 several prototypes had such cabins, including the General Aviation ST.25 (later called the GAL.41), the Vickers type 432, and the Vickers Wellingtons V and VI.

- **The GAL 41** The GAL 41 developed from the GA Monospar, and was built in 1939 to test pressurisation. Max speed, 142mph. Rate of climb, 800fpm to 16.000ft.
- **The Vickers 432** Max speed, 380mph at 15,000ft. Rate of climb, 2,750fpm. Service ceiling, 38,000ft.
- **Hawker Typhoon** First flight Feb. 1940. Maximum speed, 412mph at 19,000ft. Rate of climb, 2,740fpm. Service ceiling, 35,200ft.

In October 1939 General Aviation submitted the GAL 46 as a private venture twin-engined two-seat high-altitude fighter–bomber, but it was never regarded as a serious viable proposition.[6] However, by July 1940 the Air Ministry's interest had hardened to the point of issuing F4/40 to five manufacturers: Fairey Aviation, General Aircraft, Hawker Aircraft, Vickers Armstrong and Westland Aircraft. The specification called for a two-seat high-speed aircraft able to operate ultimately at 45,000ft and with a pressure cabin able to maintain a 10,000ft altitude equivalent. Armament was to be six forward-firing 20mm cannon, interception radar was to be carried and a maximum speed of 450mph was to be achievable at 25,000ft. Preferred power units were the Rolls-Royce RM.65 Merlin engines (the forerunners of the Merlin 60 series with two-speed, two-stage superchargers). Fuel capacity sufficient for two hours' economic cruise at 25,000ft in addition to 30 minutes at maximum power settings was required. Fuel was to be carried in self-sealing tanks, and armoured protection was required for the crew. Only two weeks was given for the five companies to submit! Vickers Armstrong and Fairey Aviation were too busy to propose designs, so the choice was between the Hawker P.1004 (a larger version of the Typhoon, with six cannons in the wings), a revamped version of the GAL 46 (which had become a pure fighter) and two designs from Westland.

The Westland tender submitted by Petter had all the signs of a classical author in style and content. It started, 'This proposed design of a twin-engined fighter, described and illustrated in this book, is put forward as being that which, after considerations of a number of types, best meets the majority of the specification requirements.'[12]

After studying the proposals, the ministry issued a revised specification, F7/41. This required a minimum speed of 415mph at 33,000ft and a service ceiling of 42,000ft. This was 60mph faster than the Spitfire and the ceiling 5,000ft higher. Significantly the top speed at high altitude was equivalent to a Mach number of 0.62, although the importance of this was not fully grasped at the time, certainly not by Petter. He could be relied upon to propose an innovative design.

The first P14, a low-drag aircraft, featured twin tandem-mounted Merlins in the fuselage; the front engine was the lower one and the second engine was mounted behind and slightly above the first. These two engines extended a third of the fuselage length, leaving the pilot's cabin midway along. The engines drove two counter-rotating propellers through a single reduction gear box (later adopted by the naval Wyvern). The six cannon were mounted in a 58ft-span wing, the crew sat back to back, and Petter ignored the requirement for a nose-wheel undercarriage, designing a conventional tail-wheel. In fact he would have been unable to find space for an undercarriage in the nose, which was occupied by the engines.

The second submission was essentially a larger Whirlwind, and he honestly described it as a Whirlwind development in the tender: 'this is a logical development of the successful "Whirlwind" and consequently can be built and tested in the minimum of time.'[3] The engine locations, the fuselage and the tailplane/fin were certainly similar. Two Rolls-Royce Merlin 61 two-stage supercharged engines would be used instead of the obsolete Peregrines. Coolant radiators would still be incorporated in the main wing leading edge. The major difference lay in the very-high-aspect-ratio wing, which had a 60ft span compared to 45ft on the Whirlwind. As a result the wing area nearly doubled and, since the take-off weight was nearly the same, at about 10,000lb, the wing loading was brought down from 40 to 20lb/sq. ft. This reduction in wing loading and the high aspect ratio was necessary to reduce the induced drag at very high altitude. The wing was also mounted mid-fuselage, instead of below it, to further reduce drag, and the engine nacelles were mounted mid-wing ahead of the wing box. The tailplane was mounted fairly low on the fin and incorporated a fairing, which this time was tapered behind the fin rather than being a forward-looking 'acorn'.

The pilot in this high-flying fighter needed a good forward and downward view, so the cockpit was much further forward than the Whirlwind's, and had a shorter nose. This left no room for the cannons in the nose, so they were first of all mounted in the wing's leading edge, but outboard of the propeller tips. They were later mounted in a ventral gun bay at the bottom of the fuselage, below the cockpit. Ammunition was stored directly behind the pressure cabin bulkhead and fed down using the Chatelleraut feed system. The two crew members were seated back to back in an armoured pressure cabin with bulletproof windscreens and a jettisonable top cover. The maximum speed was calculated to be 395mph, some 30mph slower than the other Westland submission.

The ministry also wanted the P14 to manouvre at a lower altitude and called for a 9G ultimate load factor. The increased bending moment on the wing made a stronger design necessary, with a consequent weight penalty. Petter therefore selected a thicker wing section than normal. The aerofoil section was the same as the Whirlwind (NACA 230 series). What was not fully understood initially was the compressibility effects at higher Mach

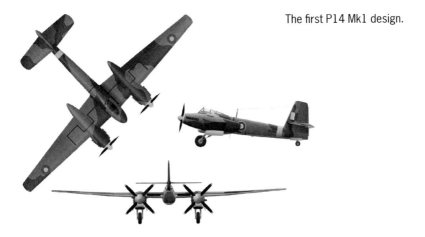

The first P14 Mk1 design.

numbers, whereby the maximum flow speed over the wing upper surface could become sonic, creating a shock wave, which triggered boundary layer separation, and a turbulent wake. The NACA 230 series had the point of maximum thickness forward on the chord, and the P14 design increased the thickness/chord ratio compared with the Whirlwind, from 17 per cent to 20 per cent at the root, and from 8 per cent to 12 per cent at the wing tip.

At this time aerofoil tests were being conducted in the High Speed Tunnels at the National Physics Laboratory, Teddington. They showed that the selected wing would have problems of high drag, poor lift and uneven aileron control, but this information was not available to Petter until later. The ministry personnel should have been aware of these tests when they approved the design. On the positive side, the scientists at the other RAE site in Farnborough had come up with a solution to the wing stalling on the Whirlwind, which was caused by the sharp leading edges of the lips of the cooling entry ducts along the wing leading edge. They suggested extending the upper lip forward of the lower one and giving it a larger radius.

The Air Ministry assesments began on 17 October and considered not just the ability to meet the specification, but also the capability of the company to turn a design into effective prototype hardware and then quantity production. General Aviation was thefore ruled out because of its lack of experience and facilities allied to the underdeveloped nature of its proposal. Hawker's design office was already overloaded with Hurricane development. Thus all of the Ministry of Aircraft Production (MAP) attention was focussed on Westland, whose design office was ready to take on new work. Whilst the ministry was judging proposals, Westland suggested a reduction in armament to four cannon but each with more rounds, and in mid-December suggested provision for just one pilot. Inevitably only Westland's proposals were considered feasible. Petter could be relied upon to produce innovation, but the low-drag, two fuselage-mounted Merlins installation was considered too complex, and it badly obstructed the pilot's forward view and needed much more development work. It was rejected. However, the use of a pressurised cabin was crucial and here Westland had backed a winner. Bill Widgery, chief experimental engineer, had designed and tested a pressure cabin control valve in the form of a 'leaking altimeter', that is it needed no pilot action. It had been tried out in a Spitfire by Harald Penrose.

Meanwhile Westland's own wind tunnels showed that the mid-wing was more efficient than the low one and the tailplane was satisfactorily clear of the turbulent wake from the flaps. In a letter to Petter on the 23 December W.S. Farren, director of technical development at the Ministry of Aircraft Production, indicated that Westland's high-altitude fighter would soon get the go-ahead. Finally, on 9 January 1941, Eric Mensforth received instructions from Lord Beaverbrook to proceed with the design and production of two P14 prototypes, the contract being for £175,000. At the design conference on 13 February, a number of changes to F4/40 were made: reduction of crew to just one pilot, and the cabin pressurisation less exacting, with a differential pressure of 3.5psi. This meant cabin altitude equivalence raised to 25,000ft at 45,000ft. The maximum speed was reduced by 10mph to 400mph. Meanwhile the designers at Yeovil had changed the wingspan from 60 to 65ft. Removal of one crew member and his equipment allowed an additional 77 gallons of fuel. The span was then increased again to 70ft and the weight increased to 18,300lb.

In March 1941 the mock-up was available for inspection, and the ministry had some criticisms, the main one being that the engines on the wing restricted the pilot's view to each side, so should be slung underneath the wings, like the Whirlwind. On 26 April specification F7/41 was issued with now more definite requirements. The P14 was to be officially named the Welkin, the poetic name for the sky or upper atmosphere.

Structurally the air frame used the metalic stressed-skin construction pioneered in the Whirlwind (except for the wooden tailplane) but this time the cockpit had light alloy armour instead of separate steel-armoured plates. It had been discovered that steel may be the best protection for projectiles striking perpendicular but for grazing aluminium became more effective.

The front pressure bulkhead was of steel, but the sides were of light alloy tapered from ½in thick; the floor was also a heavy-gauge light alloy. The top of the cabin was covered in a thick double windscreen with hot air passing between the two layers to demist. This pressure cabin was a sophisticated substructure with a pneumatically activated rubber gasket around the edges. Petter could no longer pass the rod and cable controls through the pressure cabin bulkheads without risking leakage. Instead he used electrically activated trim tabs, another first.

The air for pressurisation came from a small intake on the leading edge of the outboard starboard wing, ducted to a blower on the starboard engine, and thence ducted through a filter and silencer to a non-return valve on the cabin's rear bulkhead. This new valve automatically functioned in accordance with the actual altitude of the aircraft. It had to regulate not only pressure but temperature and humidity as well. This unit was so successful it became the basis for forming the separate Normalair Company, who supplied air-conditioning to a large number of companies worldwide. This refined system was not available in time for the two prototypes so the pilot had to suffer high temperatures from the uncooled engine compressers. Penrose reported that pilots landed soaking with perspiration, and on one occasion he caught pneumonia after leaving the cockpit in a very cold wind.

The large fuselage structure was almost identical in form to the Whirlwind, and now the rear fuselage carried a vertical 79-gallon fuel tank attached to a strengthened frame to which the tailplane was attached. The wing construction was the same as the Whirlwind except for the main spar and rear spar webs, which were a Warren-type girder instead of the plate web with holes for the air ducting.

The first flight, on 1 November, of the world's biggest single-seat fighter, was not an unqualified success, according to Penrose. 'The usual familiarisation straights were made to judge the effectiveness of the rudder and elevator. From the cockpit the wing span certainly seemed imposing.' He went on:

Then, with the hood shut, the as yet un-named Welkin was taxied to the east end, turned, and took off in a remarkably short distance; but she was very tail-heavy, requiring a strong push to hold the control column forward whilst I operated the elevator trimming-tab electric switch to remove the load. When this was eased it was immediately apparent that the ailerons were so over-balanced that the stick would have flipped to one side had I taken my hand

Harald Penrose described the view of the high-aspect-ratio wing from the cockpit as 'imposing'.

off the control column. On tentatively trying the rudder it was obviously heavy, and on deflecting and releasing, it continued to oscillate, causing a rapid hunting motion of the aircraft. This ten minute flight was restricted to one circuit.[4]

Penrose told Petter of all these problems but they were not passed on to Mensforth, who offered his congratulations.

Both the ailerons and rudder had horn balances (see Appendix 1), which were over-balancing and needed to be reduced in area, whereas the elevator was under-balanced and needed to have a horn added. Flight tests proceeded with modified control surfaces but on the fourth flight the speed control unit on the de Havilland port propeller failed, and with the rpm racing away an emergency landing was made. This was the first of a series of failures that were to plague the Welkin flight development programme.

Difficulties at the top speed end were expected when the RAE high-speed tunnel tests became available, indicating an increase in drag, and a decrease in lift near the top speed. Compressibility effects were only just beginning to affect the climb and level speed performance, although there was some deterioration in the handling qualities as well. To quote Penrose:

In speed runs at the ceiling the wings and fuselage sometimes shook as though the machine was bumping over cobblestones. My anxiety was that this might upset accurate aiming of the cannons, but I found an even more lethal snag when diving. Several dives were made quite smoothly, but on the fastest there was a sudden preliminary vibration and then the wings began to flap with increasing amplitude to what seemed several feet at the tips. Back came the throttles and gently back with the stick to slow down before a catastrophe happened – then equally suddenly the vibration stopped.[4]

What had happened was the forming of shocks on the wings' upper surface near the point of maximum thickness. These caused separation of the boundary layer, which in turn produced a turbulent wake flowing over the tailplane and elevator surfaces. This effect was known to pilots of other high-speed fighters but was occuring at a lower Mach number for the Welkin, owing to its thicker wing and the more forward location of the maximum thickness.

Petter was reluctant to believe Penrose or accept that the wing configuration would not be acceptable for high speed at altitude. The solution was either to lower the thickness/chord ratio (a complete redesign) or to extend the leading edge of the wing further forward of the front spar whilst retaining the same thickness at the spar. A fringe benefit would then be an increase in the wing area and a reduced need to use higher wing incidence to increase the ceiling a little. Petter then made a proposal to the ministry to extend the leading edge forward by 12 per cent of the chord, thus reducing the thickness/chord ratio, outboard of the nacelles, from 18 per cent down to a more reasonable 16 per cent. The Controller of Research and Development at the Ministry of Aircraft Production wrote in January 1943 that 'Petter, now that he has grown up, looks like turning into one of our best designers'. However, the response to Westland's modifications was that this improvement was not required. This was the first indication that the RAF was losing interest for a special high-altitude fighter. They may have doubted the new wing design. The ailerons produced a poor rate of roll. The twin-engined long-wingspan aircraft had a large rolling inertia anyway, so needed maximum aileron-induced forces, but this certainly reduced the manoeuvrability. The Welkin's poor response became apparent in a mock combat with a Mosquito Mk IX at 31,000ft. It was outmanoeuvred by a single-engined fighter, and making a high-speed dive was impossible. If the Merlin Mk 70 engines were replaced by the more powerful Mk 100, there is little doubt that the Welkin could have met the high-altitude maximum-speed specification. The ministry also noted that as 1943 approached the Germans showed no signs of producing a high-altitude bomber in numbers, so the initial order for 100 Welkins proceeded until about eighty had been built, and most went into store for a future eventuality that never materialised.

Petter and Davenport did not give up, proposing several initiatives to the Air Ministry. Petter proposed a two-seat radar-equipped night fighter and another with a laminar flow wing powered by Rolls-Royce Griffon engines, giving it a 50,000ft ceiling and a 425mph top speed at 40,000ft.[6] Other proposals included the use of a V tail to improve handling capabilities. He even liaised with George Carter at Gloster Aircraft to see if there was a possibility of a jet Welkin.

Only the radar-equipped night fighter was built with the blessing of the Ministry of Aircraft Production. The front fuselage was extended by 31in to provide space for an additional cockpit to house the two crew. The performance of this Welkin Mk II was disappointing, with a top speed of 333mph at 40,000ft being 50mph slower than the single seater (it was 2,000lb heavier). A second prototype was ordered to evaluate engine trials in which liquid oxygen, carried in an insulated stainless-steel tank, was injected into a specially modified Merlin engine to provide more power at speed and altitude. The hazards of handling and pumping LOX on the ground and in the air prompted the abandoning of this development.

The Welkin Mk II survived for a number of years after the Second World War, firstly flown by Westland in a pressure cabin development programme, and then for the Ministry of Supply for further radar trials. In 1946 it made its public debut when Penrose flew it at the first SBAC air show. It seems that the prime legacy was the development of the pressurisation system and launching of Normalair as part of Westland Technologies Division. Another legacy was learning the hard way of the perils of compressibility when flying fast and high with unsuitable wings.

4

THE ENGLISH ELECTRIC YEARS: THE CANBERRA, 1944–1950

By the end of 1944 Petter's reputation with Westland's main customers, the Air Ministry and the Ministry of Aircraft production, was probably high. In spite of his occasional outbursts against governmental officialdom, his status as a designer and chief engineer had grown steadily with time. The Lysander was a technical success. The innovative Whirlwind was also a technical success, falling short of scheduled delivery times solely because of its Rolls-Royce engines and superchargers. The Welkin was not a success because of the compressibility problems with its thick wings, but Petter's proposals for a longer wing chord extension would probably have been successful had the German threat of high-altitude bombers materialised.

It should also be noted that Westland had produced other aircraft under license and in large numbers. About 1,600 Spitfires had been produced, and later Seafires. Consequently Petter was well thought of by Sir Wilfred Freeman, chief executive officer at the Ministry of Aircraft Production, and by N.E. Rowe (known by everyone as 'Nero') the current director of technical development. Discussions between the three of them led to the concept of a twin jet-propelled fighter–bomber to replace the Mosquito. Specification B1/44 was issued by the ministry for an aircraft with an all-up-weight of 32,000lb, a payload of 4,000lb, a top speed of 530mph, manoeuvrability at low altitude (in the same fashion as the Mosquito) and powered by two engines with a combined thrust of 8,000lb. This last requirement was particularly adventurous, since the first jet to see service was the Meteor in July 1944, having Rolls-Royce Welland engines of

1,600lb thrust. Rolls-Royce had meanwhile designed and produced prototypes of the Nene, with a static thrust of 5,000lb. All these engines followed the original design of Whittle in 1941, using a centrifugal compressor. This involved taking in the air along the central axis, turning it through a right angle and using its increasing speed with radius to compress and concentrate the flow before directing it into the combustion chambers. These engines therefore had a large diameter and frontal area, which grew with each new development, and therefore increased the total drag. This posed a challenge for any aircraft designer wishing to control the drag of a high-speed fighter–bomber.

Petter was one such designer, and he rose to the task, even persuading Westland to put up some private venture capital and manufacture a mock-up of a possible forward fuselage with engine intakes and ducts. As usual his enthusiasm for new aircraft and new problems was undiminished, and he proposed a design owing something to his first low-drag submission for the Welkin. It involved two engines in the centre of the fuselage, fed by ducts passing underneath the pilot to intakes low in the nose. He realised this configuration created problems with finding room for the bomb bay,

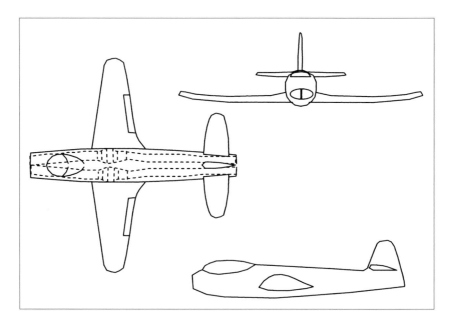

Petter's twin jet design of March 1944.

which had to be located near the centre of gravity. Room also had to be found near the centre of gravity for the fuel tanks, to avoid trimming problems. Petter was aware of these conflicts as the design progressed, and had embryonic ideas for engines in nacelle installations.

Throughout his career Teddy was known to suddenly take off from his work for a period of up to six weeks or so, without warning, usually at a time of stress or frustration. He was a deeply religious man and it was believed he fled to a monastery or religious commune, to return refreshed and bristling with ideas. These days it might be thought he was bipolar, a condition not uncommon in people of genius in the sciences and the arts. In April 1944 he suddenly left his work, and rumour had it he'd travelled to Switzerland, a refuge to which he would return towards the end of his career.

Whilst he was absent, Mensforth decided to put the project office to work on a specification N11/44 for a naval single-seat fighter of 23,000lb all-up-weight, to be powered by a Rolls-Royce turboprop unit driving contra-rotating propellers. This aircraft was to have 45ft folding wings, and a very ambitious pay load, range and high-/low-speed performance (it was later to become the Wyvern). All work on the jet bomber was stopped when Mensforth succeeded in getting a go-ahead from the Navy to proceed with a design. When Petter returned he was furious, because he knew that Westland did not have the resources to build both aircraft. He believed that the top management had connived in his absence to eliminate his pet project. He handed in his immediate resignation.

There must have been some stormy meetings between Petter, Mensforth and the John Brown Board, but he eventually succeeded in obtaining the rights to the jet bomber when he left Westland Aircraft. He was replaced as technical director by Arthur Davenport, whose position as chief designer was taken by John Disley.

Petter undoubtedly turned to Sir Wilfred Freeman at the Ministry of Aircraft Production, with whom he had established a respectful relationship. He might also have visited Sir Ralph Sorely; in any event he was introduced to Sir George Nelson, the chairman of the English Electric board.

English Electric was formed in 1918 as an amalgamation of five engineering firms at Battersea, Kingston, Coventry, Brough and Preston. It made flying boats in the First World War, designed and built some Kingston naval biplanes, and the Wren motor glider in 1923, but only three were sold. In 1926 English Electric ceased trading in aircraft.

After twelve years of dormancy, the aircraft division was reformed in 1938, when the government initiated its rearmament programme. A production line was set up at Strand Road, Preston, believed to be relatively safe from enemy attack, where it had a large and skilled workforce available from its tramcar fabrication operations. Halifax bombers were eventually produced at the astonishing rate of one a day in 1942, and the factory eventually built more than a third of all Halifaxes in Britain. Vampires were produced at the rate of four per week; in fact English Electric built more Vampires than did de Havilland.

This successful manufacturing company was built up by Arthur Sheffield, who was a close associate of Sir George Nelson. He was in charge of Hampden and Halifax production at Preston and Samlesbury. He was absolute monarch and intolerant of outside interference. He was, if possible, even more fanatically devoted to the welfare of his factory than Petter was to aircraft design and development. 'Sheff' owed his position and status to Sir George, to whom he was loyal in the extreme. He would only take instructions from him, ruthlessly and sometimes misinterpreted. For example Sir George had indicated that an increase in the production of diesel engines was needed to meet demand. The next day, key aircraft components were taken off several machines and replaced by diesel engine parts without consultation.

Sir George saw that in 1945 the tremendous pace of aircraft development was poised to continue, with both military and civil aircraft using the powerful engines which had become available. He may have thought that some wartime designs were unsuccessful or even dangerous, and some factories had a poor reputation, so there could be an opportunity to both design and build new aircraft. It is also possible that he was encouraged by politicians who believed that decentralisation from London was desirable, but Sir George was aware that he needed talent to break into the established circle of experienced aircraft design firms.

Petter would have been introduced to the English Electric board as an experienced designer with an aircraft already designed as a replacement for the Mosquito fighter–bomber. He formally joined English Electric in July 1944 as chief engineer of the aircraft division at Preston. As chief engineer, and not just chief designer, he could reasonably expect to have control over the experimental workshop at Strand Road. Petter believed that all experimental facilities and prototype production should be under the control of the chief engineer, completely separate from the production line with its tight rules and controls to ensure production quality and reliability. Future conflicts between Petter and Arthur Sheffield seemed assured.

Meanwhile Petter set about the task of recruiting the best possible design team. It was to be located in a pre-war motor showroom (Barton Motors) that had been converted to a government training centre, known as TC. It was situated near Preston railway station, only about a mile from the main works at Strand Road. The first floor had been converted to offices for the senior staff, aerodynamics, stress and design. The ground-floor showrooms were for weights, lofting, mechanical tests and progress departments. Behind these, the mock-up and experimental facilities were built. Working conditions were primitive, so prospective recruits were assured that this was only a temporary location, but for most of the design team it lasted for three years.

The most important recruit was Frederick Page, who met Petter in October 1944, when he was senior aerodynamicist at Hawker. He

Petter at his desk in the T.C.

had been responsible for the successful design of the spring tab controls (see Appendix 1) for the Tempest fighter which had been so manoeuvrable in defence against the V1 flying bomb. The Tempest fought battles in Germany with speeds above 500mph. Petter offered Page the position of chief stressman. He discussed with him his early ideas of a replacement for the Mosquito, using one centrifugal engine based on the Nene and mounted in the fuselage. Page agreed with Petter's reservations that the engines ahead of the bomb bay and fuel tanks would make it difficult to balance the aircraft. He seemed to have told Petter directly that 'Westland were not in the same league as Hawker, and the Whirlwind and Welkin had not been successful. The Lysander was a competent but unexciting army co-operation aircraft.'[9] Nevertheless, an offer was made to Page, who accepted it under specified conditions. As well as stressing he wanted to be in charge of weight balancing, all drawings would have to be routed through the stress office and signed. He would also be in charge of structural and mechanical testing. In view of his experience at Hawker, it was agreed informally that he could make suggestions on aerodynamic design. Even though he was only 27 years old (nine years younger than Petter) he was clearly thought to be a significant asset for the design team. Sydney Camm at Hawker had apparently advised him that 'You must be mad to leave Hawker. Nothing happens north of the Thames.'[3] Page received his letter of appointment in April 1945.

In 1945, with Petter on board, English Electric was placed on a shortlist of aircraft companies with Avro, Handley Page, Short Bros and Vickers to proceed with jet bomber design studies. As it turned out, all these companies, with the exception of English Electric, concentrated on the larger V-bomber designs with four engines, so leaving the field clear for English Electric on the twin-engined E3/45 which soon became the B3/45 specification.

Recruiting people to work in Lancashire was not easy because of reluctance to change and the perceived cultural shock moving to an area of widespread poverty. The Control of Employment Act meant that recruitment was restricted to companies that already had contracts to design and produce. A catch-22 situation. Whitehall philosophy, as opposed to the ministries, was that no new military aircraft were needed for ten years. Petter surmounted this restriction by proposing a study contract for investigating designs for a high-speed bomber. He clearly did a good sales job and received a contract in June 1945, with an expenditure of £1,000. He had a foot in the door.

His next recruitments were people he knew personally. Harry Harrison came from Westland to become chief draughtsman. Dennis Smith became assistant to the chief engineer. Don Crowe was recruited from Handley Page, for whom he had been liaison engineer for the build of Hampden and Halifax bombers at Preston, and now became Petter's chief structural engineer. (In fact by the end of 1945 all Halifaxes coming off the production line at Strand Road were moved to the other side of the airfield to be cut up as scrap. The crew seat cushions were salvaged for the offices at TC).

The position of chief of aerodynamics proved to be a problem for nearly a year. Petter's experience with the Welkin had made him realise the importance of compressibility for any new high-speed bomber. He approached Elfyn Richards, a young scientist at the National Physics Laboratory, who was reluctant to move to Lancashire. As it happened the chief engineer at Vickers in Weybridge was looking for an opportunity to get Richards to join Vickers, so he suggested to Petter that he should employ his present chief of aerodynamics, Dai Ellis, who was more of a general mechanical engineer anyway. This he did. Eventually, in mid-1946, Ray Creasey followed Ellis from Vickers to take charge of aerodynamics. Ray Creasey turned out to be a brilliant and imaginative aerodynamicist, whose skills were crucial for the P1 and Lightning, but who unfortunately died prematurely in 1976.

During his first year at English Electric, and almost alone, Petter had looked at possible designs for the Ministry of Supply specification B3/45, the key elements of which were:

- a high-speed, high-altitude unarmed bomber
- two engines, and capable of taking off should one fail
- maximum speed at 40,000ft to be greater than 500mph, and service ceiling to be not less than 50,000ft
- crew of two, pilot and radar operator/navigator, in a pressurised cabin
- full load take-off in 1,400yds and landing in 1,000yds.

His early ideas, brought from Yeovil, had used single and twin engines located in the fuselage, but in consultation with Page he had changed to two engines mounted in the thickest part of the wing, next to the fuselage. The wings were to have a 300° sweep. This had become an option after Petter and Page visited the Rolls-Royce factory at Barnoldswick, where Stanley Hooker was developing the AJ 65 axial flow engine.[10] These would

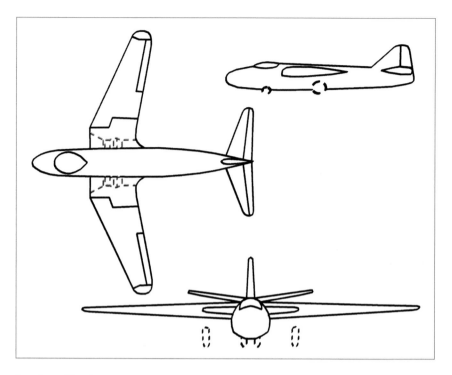

A projected bomber design in 1945 with engines embedded in a thick swept wing.

have much smaller diameters and a thrust of 6,500lb, so two would satisfy the requirements. Rolls-Royce was experiencing some aerodynamic surge instabilities, but expected to overcome this problem; this they did and the engine became the Avon, a very successful design for years to come.

Contemporary Design Alternatives

The idea of using swept wings to overcome compressibility problems was current at the time, and was adopted by several competing designs in France and the US.

> **Sud–Ouest Vautor** First flew Oct. 1952. Sweep, 35°. Low-level bomber. Max speed, 687mph at sea level. Rate of climb, 11,200fpm. Service ceiling, 50,000ft. Wing loading, 82lb/sq. ft. The French Air Force took 140 of these aircraft (the last in 1979) but the only other customer was Israel, who bought 28 aircraft. It lost out to the superior Mirage III.

Sud-Ouest Vautor.

Martin XB -51.

Martin XB-51 First flew in 1949. Swept wings at 35°. Three engines. Low-level bomber. Max speed, 645mph. Service ceiling, 40,000ft. Rate of climb, 6,980fpm! Wing loading, 72lb/sq. ft. This aircraft had an idiosyncratic design, with three engines and clean wings with a host of innovative controls. However, its performance and manoeuvrability were poor, and ironically it lost out to the Canberra, built under licence by Martins.

Boeing B 47 First flew Dec. 1947. Max speed, 607mph. Rate of climb, 4,660fpm. Service ceiling, 33,100ft. Wing loading, 93lb/sq. ft. This aircraft had thin swept wings. The design team solved the engine problem by having three on each wing, mounted in pods slung underneath. Its drag was somewhat high but the concept was truly revolutionary and was adopted later by the B 52 long-range bomber, and virtually all civil aircraft since. The B 47 was designed to deliver nuclear weapons to the USSR at altitude. It was not seriously manoeuvrable.

Boeing B 47.

Other competitive designs also used unswept wings:

- **Ilyushin Il-28/Hong H-5** First flew 1948. First USSR aircraft to
  enter large-scale production. Max speed, 560mph at 14,750ft. Service
  ceiling, 40,850ft. Rate of climb, 2,950fpm. Wing loading, 60lb/sq. ft.
  This aircraft was a strategic and economic success for Russia: more than
  6,000 were built and many exported. Its original engines were Nenes,
  later replaced with Russian copies, delivering 6,000lb thrust. Its wings
  were quite thin, and the wing loading modest, so its performance
  without using sweep was creditable.
- **North American Aviation Tornado B45** First flew in March 1947.
  Max speed 570mph. Rate of climb, 5,880fpm, Service ceiling, 46,400ft.

74

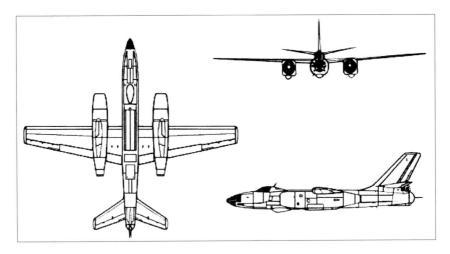

Ilyushin 11-28/Hong H-5.

American Tornado B45.

Wing loading, 72lb/sq. ft. This aircraft was the first US jet bomber, and 143 were built. It was an important part of the US Strategic Air Command in the early fifties. It had a high–aspect–ratio, thinnish wing and needed four engines to achieve its speed and altitude. It was a very large bomber, weighing more than 81,000lb. The total thrust from the four engines exceeded 20,000lb. This explains how it managed to achieve its respectable performance. It was eventually displaced by the B 47 Stratojet, which had swept wings.

The Development of the Canberra Design

By September 1945 the basic English Electric A1 design was fixed. Petter and Page had agreed that the two engines should sit in nacelles on the wings, but there remained the choice of wing thickness and sweep angle. Petter had previously experienced compressibility problems with the Welkin,

Canberra Configuration.

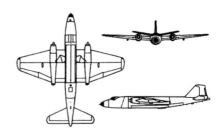

which had a wing of thickness/chord ratio of 22 per cent at the root, and a critical Mach number of about 0.6. His solution was now characteristically ingenious. He scorned the loss of performance due to sweep, and went to an unswept wing of constant chord between fuselage and nacelles and a tapered section outboard. The wing section was an uncambered NACA 01240 inboard (t/c ratio, 12 per cent) and uncambered NACA 0840 (t/c ratio, 8 per cent) outboard. The wing root chord was long, 19ft, giving a maximum root thickness of more than 2ft, which was easily adequate to house the main landing gear.

The design looked distinctly old-fashioned compared to other current designs, but the most outstanding feature was the wing loading of 48lb/sq. ft, which was much lower than any of the other current designs (listed in the table). The other important parameter was the thrust/weight ratio, which affects the rate of climb, manoeuvrability, etc. Most of the aircraft have a similar value near 0.25, but the Canberra's was 0.32.

	Sud–Oeust Vautor	Martin XB–51	B 47	Ilyushin II–28	B 45 Tornado	Canberra
Wing loading (lb/sq. ft)	82	72	93	60	72	48
Thrust/ weight ratio	0.4	0.28	0.22	0.25	0.26	0.32

The low value of wing loading at 48lb/sq. ft meant that the aircraft could operate at quite modest angles of incidence, with corresponding lower flow speeds over the uncambered wing, and thus achieving a maximum speed of 580mph (Mach > 0.8 at 40,000ft) without a local supersonic region. The other significant metric is the thrust/weight ratio, which together with the low wing loading gave the Canberra the manoeuvrability of a fighter.

Petter is on record as saying that 'sweep was hardly worth it, increasing the critical Mach number only from 0.8 to 0.9'.[8] The choice of a low wing loading also brought with it a host of additional advantages.

First, the wings had a low aspect ratio of 4.3, which made for a lightweight structure but usually a high induced drag (see Appendix 1). However, because the wing operated at such a low incidence, the induced drag was tolerable.

Second, the wing surfaces had very modest curvature, so the intersections with the fuselage and the nacelles needed no fillets to smooth the flow.

Third, the low-aspect-ratio wing in roll had both low aerodynamic damping and low rotation inertia. The aircraft manoeuvrability was consequently better than any fighters that might attempt to intercept it at altitude. It also meant that the Canberra became a great success at low altitude when the specification was changed.

Details of the A1 design had been kept fairly confidential, but the existence of the project was officially revealed when George Strauss, Minister of Supply, made a statement about the project, including the comment that 'it would have about twice the speed of current bombers at altitude'.

In October 1945 the Ministry of Supply issued an updated specification for the B3/45, to which English Electric responded immediately. Consequently the ministry issued a contract in January 1946 for the detailed design and manufacture of four prototypes. A key feature was a specification for a radar bomb-aiming system, so that a crew of only two, a pilot and navigator, would be required.

With many engineers, designers and draughtsmen now on board, the facilities at the ex-training centre were becoming strained. The cellar was brought into use for a small wind tunnel. A mock-up of the front fuselage was started, in order for the team to plan the layout of the operational equipment and the crew's seating locations. Room was also found for the full-scale lofting, in which the lines for the B3/45 could be laid out. By mid-1947 the total design team exceeded a hundred, so permission was given to take over the hangers and runway at Warton Aerodrome, an ex-USAAF base some 10 miles west of Preston. This would enable the development of the prototypes and new equipment to be separate from the production lines at Samlesbury, and kept Petter at arm's length from Arthur Sheffield, who resented the separation of research and development from production.

The basic detailed design of the Canberra was now fixed. Comparing this final design with the one Petter brought from Westland, the gross weight had increased from less than 30,000lb to 40,000lb, largely due to scaling up the size to carry a larger bomb load. Petter's philosophy was still to keep it simple and opt for minimum drag if possible. Thus the wings were mounted mid fuselage and the engines mounted at the wing centre line. The fuselage was a long cylinder of circular cross section with a gentle forward droop in front of the pilot, to give a first-class view, and an even more gentle upward taper to the tail unit to give a ground clearance for steep landing (the undercarriage, which retracted into the wing, was quite short). The overall view was of simplicity and elegance, and was often quoted as the most handsome ever for a bomber.

A problem of locating the Avon engines, small diameter though they had, was that the jet pipe had to pass through the main wing spar. Petter's solution was to use an alternative to built-up sections, namely light alloy forgings suitably machined and drilled. He had first used large forgings at Westland and would use even more on the future Gnat. In this case he used a 'banjo' type forging to carry the spar loads around the jet pipe.

The 'most handsome of bombers'.

The spar booms continued from this forging up to the fuselage frame, another forging, to which the spar booms were bolted.

The engines were mounted in front of the main wing spar, its nacelle being two ribs cantilevered forward and supporting a tunnel roof structure from which to hang the engines. This roof structure was a series of closely spaced semi-circular frames, able to take the bending moments and shear, but open at the diameter to enable the engines to be lowered.

In addition to the wing's main spar, which was at right angles to the fuselage and extended span-wise and continuous to the tip, there was also a rear spar to take the loads from the flaps and ailerons, and which followed a more circuitous route to the tip rib.

As part of Petter's relentless pursuit of aerodynamic cleanliness, the wing leading edge was formed by a single sheet of light alloy wrapped around way back to the 40 per cent chord position, and supported by Redux-bonded stiffeners and ribs normal to the leading edge. The ribs were not attached

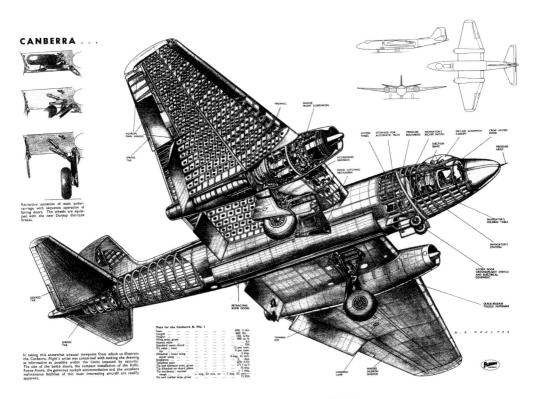

View showing main features of wing structure. (Courtesy of *Flight*)

directly to the wing surface but passed underneath the stiffeners, to which they were attached by miniature eye bolts and threaded anchor cleats. These small screw adjusters were used to obtain a highly accurate wing profile in a jig against external contour boards. The wing profile was therefore very accurate back to the main spar, having no external joints or excrescences, with the aim of maintaining laminar flow in the boundary layer, and no transition leading to separated flow. It was not certain, though, that this could be achieved, since wind tunnels were not large enough to simulate flow at the correct Reynolds number (see Appendix 1). Also, in service the wing surface would become contaminated by scratches, dust and insects.

The leading edge cell, outboard of the nacelle, contained an integral fuel tank. (Three other tanks were in the fuselage.) The leading edge cell between the nacelle and fuselage housed the retracted undercarriage, mounted from the main spar, and many of the accessories driven by the engine, such as a Rotol/Rolls-Royce auxiliary gear box driving a generator, a Lockheed pump and, on the starboard side, a Hymatic compressor. On the port side a Marshall cabin blower was carried instead of the compressor. The mixing valves and cooler for the pressurised cabin air-conditioning system were also in the port side.

The wing main spar supported split flaps, inboard between nacelles and fuselage, and outboard up to the ailerons. Each flap was actuated by a hydraulic jack set span-wise along the face of the spar web.

The leading edge of the ailerons was contained within a shroud that was cantilevered from the rear spar. The 'beak' nose balance on the aileron was of the Westland type, but did not use a flexible diaphragm between the nose and the rear spar (see Appendix 1). Instead the clearance between the balance spine and the spar web had been reduced to less than 1.5mm. To achieve this tolerance the shrouds were jig built. Indeed the English Electric practice was now to use jig building for the complete aircraft to

Anchor cleats for adjusting rib web edge to stiffener lips.

ensure an accurate structure. No experimental shop existed to manufacture the first prototype; the pre-production jigs were designed for the first twenty aircraft or more. They would provide experience and useful data for final production jigs, should a large number of aircraft be ordered.

The Canberra fuselage gave the appearance of being especially long and slim, although in fact the diameter of the central portions was 6ft. The fuselage was supported by light circular channel frames, which were notched to take the longitudinal extruded 'Z' stiffeners (to prevent the thin fuselage skin from buckling). The wing structure did not pass through the fuselage in the traditional way, but was bolted, at both main and rear spar, to substantial forged frames instead of bulkheads, leaving the interior space free. However, a full bulkhead was used behind the cockpit to withstand the cabin pressure and also support the rear-folding nose wheel assembly.

The pilot's canopy was a development of the pressure cabin built for the Welkin. It had a double Perspex dome in which the outer skin was separated from the inner by a three-eighths of an inch cavity aspirated through twin packs of activated alumina desiccant. It was anchored to a foundation tube by means of explosive bolts which could be jettisoned in an emergency. A small two-ply glass screen was mounted immediately in front of the pilot's face as a protection between the canopy being jettisoned and the pilot following suit. Behind the pilot, on the centre-line of the fuselage, was the navigator's seat, also equipped with a Martin-Baker ejector seat mounted on the bulkhead sloping backwards and forming the rear diaphragm of the pressure cabin.

The bomb bay, aft of the nose wheel well, was 23ft long and had sliding, rather than hinged, doors, which retracted upwards on three roller guide tracks, one at each end and one in the middle. The bomb bay roof was a massive structure since it had to support two fuel tanks. This roof was built up of longitudinal channel beams riveted to transverse beams so that a grid skeleton was formed as the floor for the fuel tanks.

About 27in aft of the bomb bay was the transport joint to the rear fuselage with its circular channel frames and extruded longitudinal Z-stringers. The channel frames eventually gave way to a pear-shaped forging, to which the fin post was attached.

Unusually for a bomber, the aircraft had a variable-incidence tailplane, electrically actuated to cope with longitudinal trim changes resulting from bombs released or fuel consumed. The tailplane structure was a scaled-down version of the main wing, and had pivot points bolted to the forward 'pear' frame.

The fin had the orthodox spar/rib construction but the leading edge cell was made of wood. The reason for this was the use of aerials embedded in the structure. The rudder was a conventional light alloy but had a wooden horn balance.

In order to reduce the stick and pedal forces to acceptable levels, very precise aerodynamic balancing was needed. There were three types available (see Appendix 1): horns, internal shrouded balances and various types of tabs, of which the spring tab was the most effective. It was decided to use horn balances on the elevators and rudder, internal balancing on the ailerons and spring tabs on all control surfaces. Wind tunnels were not yet large enough to test these features accurately, but Petter used his knowledge of the Westland Irving internal balance, and Page used his extensive experience of spring tabs at Hawker, to design these controls on the Canberra. The spring tabs would play a major role in contributing to the Canberra's high manoeuvrability over its whole speed range.

The only real deficiency in control was the use of rudder should one engine fail at take-off. This was essentially a safety problem and would lead to the addition of power control for the rudder in later aircraft.

In designing this jet-powered bomber, Petter perfected an airframe to be manufactured as a set of five independent primary structures:

- the front fuselage
- the centre fuselage
- the rear fuselage
- main-plane with non-anti-icing Avon Mk 1 engines
- main-plane with anti-icing Mk109 engines and integral fuel tanks.[11]

This flexible strategy was to prove invaluable in future years, when different roles for the aircraft were defined and put into service.

The one remaining position to be filled was that of chief test pilot. English Electric had several competent production test pilots used for testing the aircraft during the war at the Samlesbury factory airfield. None, however, had experience of flying experimental prototypes. Moreover they reported directly to Arthur Sheffield. Petter could not tolerate having his chief Test Pilot going over his head to Sheffield in areas of technical disagreement, or just routine modifications needed after a few flights. With Page's agreement he decided to look outside English Electric for the first man to fly the Canberra, and to challenge the establishment by making his (Petter's) own appointment.

The shortlist came down to two: Squadron Leader Tony Martindale and Wing Commander Roland Beamont. Martindale had achieved distinction at the Royal Aircraft Establishment in high-speed diving tests to establish absolute limits of control due to compressibility. He had survived some remarkable experiences, including landing safely after the total loss of a propeller from a Spitfire in a vertical dive at Mach 0.9. Beamont, in the years 1939–1946 had completed three tours of operations in the RAF in Typhoons and Tempest ground attack, Tempest defence against V1 flying bombs, and air fights over France and Germany from D-Day onwards. After the war he spent time on jet fighter development with the Meteor VI at Glosters, and then two spells as experimental test pilot at Hawker Aircraft. Beamont was therefore well known to Page.

Beamont had actually been interviewed by Petter six months previously, on which occasion neither man was impressed with the other. Beamont was asked to provide a reference, and suggested a very senior air marshal, who he found out many years later had advised against his suitability for the post. So Petter initially favoured Martindale, who had a degree in Engineering and could be relied upon to match practice with theory. However, because he worked at the RAE he might be expected to express technical opinions that Teddy did not need. His occasional confrontations with Penrose at Westland had developed his strict ideas on the limits to which he would allow recommendations from his test pilot to be considered. In the event, in May 1947 both Petter and Page chose to go with Beamont because they felt his extensive operational experience would be important in Canberra development and in ensuring a favourable reception from RAF serving pilots.

Over the next few years it turned out that Beamont was one of Petter's best appointments. There was only one highly charged confrontation, when Beamont challenged a decision he thought was in his area of responsibility. According to Beamont, years later, 'Petter seemed to explode and I said perhaps we had better forget our arrangement and I would write out my resignation forthwith. All right he said and I was dismissed. About half an hour later he sent for me and said we must not be too hasty.'[3] No further confrontations ever occurred. Beamont stood his ground on all matters relating to safety and operational requirements, but did not try to take over the role of technical staff, towards whom he had developed high confidence and respect. His aggressive but highly skilled approach to flight demonstrations in the UK and in the US showed great faith in the robustness of the Canberra and went a long way to persuading many potential customers of its virtues.

By the end of 1948 all the personnel had moved from the TC and Petter had established his regime for design and testing at Warton. Dai Ellis had started the design of several wind tunnels. However, all fabrication, including experimental work, still remained at Strand Road, Preston, under the control of Arthur Sheffield, who had the full support of Sir George Nelson. This would lead to a major confrontation between Petter and Sir George.

The last appointment Petter made to his senior staff was A.E. Ellison, from Airspeed Ltd, who came in as assistant chief designer in mid 1948. Don Crowe took over as chief draughtsman and H.C. Harrison became chief production designer. The design team was now complete, and being self-contained at Warton was an example of what became known as 'project management'. As the team had no previous aircraft to worry about, it was able to concentrate solely on the Canberra. It was probably the most formidable military design team in Europe, and remained so for many years, producing the Canberra, P1, Lightning, Jaguar, TSR2, Tornado, EAP and Eurofighter/Typhoon.

In late 1947 the project suffered a major setback when the Ministry of Supply revealed serious technical problems with the radar for high-altitude bomb aiming. The radar system was a development of that used in the war but enhanced for high-altitude accuracy. It was becoming apparent that the radar system was growing too large to be installed in the nose of the Canberra. The ministry stated that the radar-aimed, high-altitude bombing mission would have to be delayed and assigned it to the larger volume capacity of the V bombers, which all had later delivery schedules. However, they remained enthusiastic for the B3/45 design and possibly felt some guilt that the problems were not of English Electric origin, so they undertook to find a new mission for the aircraft. Inside English Electric there was much consternation about the future of the programme and Petter set up his own in-house team to define some concepts for other missions for the Canberra. This quickly led to three potential roles that could be submitted for consideration by the ministry:

- To reduce the altitude of operation so that manual bomb aiming could be achieved. This called for a third crew member and the range would be greatly reduced from the B3/45 specification of 1,600 miles. The

The design team, from the left: F.D. Crowe, D.I. Ellis, S.C. Harrison, A.E. Ellison, W.E.W. Petter, R.P. Beamont, D.B. Smith, F.W. Page and H.S. Howatt.

ministry accepted this inevitable range reduction and incorporated it into a new specification for a tactical bomber, the B5/47. With a bigger bomb load, the range ended up as 1,000 miles.

- Another mission role could utilise the Canberra's high-altitude capability with the improving technology of high-altitude photography. This new specification was for photographic reconnaissance and only needed a crew of two, but would need eight large cameras. The range for this role needed to be 3,000 miles, so the forward part of the bomb bay would be converted to a 400-gallon fuel tank.

- The third variant was a dual seat conversion to be used as a navigational trainer.

All these variants would later be developed for production, and the extra design and fabrication work would prove very lucrative for the company in the long run. The consequent fabrication effort caused havoc in the English Electric offices and workshops, which led to bad feelings between Petter and the works manager, Arthur Sheffield, who believed the designers must, in some way, have been responsible for getting the company into this chaotic situation.

Ironically, therefore, the Canberra would never be used for the very-high-altitude bombing missions for which it was initially designed, but the design had enough flexibility to perform a variety of missions, some of which were still undefined at this time. The singular choice of using a low wing loading to avoid local supersonic flow over the wing also gave the aircraft the ability to suddenly increase incidence and lift without stalling. This factor was an extraordinary quality for a bomber, which traditionally had poor manoeuvrability.

Additional unexpected problems were occurring at Rolls-Royce for the Avon engines. To quote Stanley Hooker, 'In 1946 we had the first AJ65 engine on test at Barnoldswick. It was difficult to start, would not accelerate, broke its first stage blades and could only be coaxed reluctantly to 5,000lb thrust. It was comparatively speaking a hell of a mess.'[10] This failure also led to panic at Rolls-Royce. Hooker was transferred to other duties and the entire compressor programme was moved from Barnoldswick to Derby, where Lord Hives would take charge. It was clear that a significant development programme was required to match the pressure rises through each stage of the compressor, and that blow-off valves may be required to operate during starting and accelerating.

At English Electric there was similar consternation that the Avon may not be ready in time. Plans were made to equip one of the four contracted Canberras with the centrifugal Nene, at least for the first proving flights. The installation of the Nene would greatly increase the nacelle diameter and certainly spoil the clean lines of the aircraft to which Teddy was so dedicated.

The issue of the aesthetics of his aircraft was always important to Petter. He would give a paper to the Royal Aeronautical Society in which he compared creative aircraft design with great art and poetry.[14] He genuinely believed in a correlation between aesthetics and engineering efficiency. If it didn't look right, it probably wasn't right.

In 1947 Dai Ellis started to design and install three new wind tunnels at Warton. The first to become operational was the low-speed tunnel, having a working section 9ft × 7ft. The aerodynamic tests of the model Canberra confirmed predictions made by Ray Creasy's group regarding stick-fixed stability and the effect of the chosen tailplane and fin. Several flap configurations were tested to see their effect on the flow over the tailplane.

The sliding bomb bay doors were tested and the loads induced by opening them assessed, together with the pressure fluctuations in the bomb bay to ensure clean bomb release.

The tunnel tests also confirmed the range of the tailplane angles needed to assist take-off and landing, and to cope with any trim changes arising in dives due to Mach number effects. An issue requiring special attention in the wind tunnel tests was the amount of rudder power needed to cope with angles of yaw. The pilot needed sufficient lateral control to compensate for engine failure at take-off. To keep the aircraft on the runway, without having an excessively large fin, it was necessary to have a rudder both powerful and well balanced for it to be effective within the leg strength of the pilot.

The objective in designing the controls for the first flight was to select flyable but safe configurations. Insufficient balance would produce stick and pedal forces that were high, but not beyond the pilot's strength. Over-balancing would lead to unstable oscillations that could be uncontrollable with the pilot's finite reaction rates.

By mid 1948 Rolls-Royce was making good progress on the Avon development, and it appeared likely that the engine would be available for the first flight. Two prototypes were achieving a thrust of 6,000lb (instead of 6,500lb) when installed in a Lancastrian test bed. It would take several more years of development at Rolls-Royce before the ministry would be satisfied with the in-flight handling of the Avon, but at least its availability would no longer be a problem in the A1 development.

In March 1949 the Ministry of Supply had such confidence in the three roles for the aircraft proposed by Petter that it placed a production order for 132 aircraft, comprising ninety B5/47s, thirty-four PR31/46s and eight T2/49s. The first flight of the Canberra (which was still to be named) was some months away, so this endorsement, or act of faith, was unusual but prompted by the increasing hostility of the Soviet Union.

Early in 1949 components were shipped to Warton from 'TC' and Strand Road. The first prototype, EEA1, serial number VN799, resplendent in 'Petter' blue, was unceremoniously rolled out of No. 25 hanger at Warton on 2 May. Engine runs were completed in the next five days. Next followed several days of taxiing 'hops' to test the ailerons, flaps, elevator and rudder. The test pilot, Roly Beamont, reported some minor nose-wheel 'shimmy',[8] and then, on 9 May, aircraft No. 799 accelerated quickly with a flap setting of 30° to give maximum lift. The first check was for elevator 'feel', and then the ailerons, and finally the rudder. The first two cases showed smooth and precise controllability, but the rudder was tested with only limited displacements, owing to the need to hold a straight line on Warton's narrow runway.

These straight 'hops' were remarkable in themselves as they confirmed the design predictions that this high-performance bomber could easily take off in less than 500yds, land again and brake to a standstill comfortably within the 1,900yd runway at Warton, without even overheating the brakes. In a mood of quiet confidence the decision was made to go for first flight as soon as the appropriate weather conditions were available. This happened on 13 May.

Traditionally the first flight is a special occasion at any aircraft factory, but it was doubtful whether any of the English Electric personnel had a

The first 'A1' prototype.

New fin with trimmed rudder horn.

full appreciation of just how momentous the occasion was. The design was the first major aircraft to originate from this Preston factory. It was Britain's first jet bomber, and the first to be designed around axial-flow jet engines. It was the first brainchild of a new team of designers, whose future clearly depended on its success, as did the future of the Preston factory, whose Vampire contract was to run out the following year. It would, in one step, more than double the performance available to RAF Bomber Command.

The weather on the day of the thirteenth was excellent. A Vampire chase aircraft was ordered on standby from 10.45 a.m. at Samlesbury, to be flown by test pilot J.W.C. Squier. In later years, as chief production test pilot, Squier flew more than 3,000 Canberra test flights. The final discussion between Petter and Beamont took place, and the pilot suggested that Friday the thirteenth was not everybody's idea of a lucky day, but he was more than ready to fly. Petter then characteristically disappeared from the scene, and reappeared more than two hours late to take charge of the post-flight debriefing.

The Avon engines started smoothly, and 799 was taxied slowly down the short runway and then to the end of the main runway. The A1 became airborne at 90mph and climbed shallowly to 138mph, Beamont tested the rudder, which gave the correct response in yaw, but then suddenly increased with a jerk and sharp reduction in pedal force. At 10,000ft the aircraft was in precise trim, and responses to small inputs to ailerons and elevator were undramatic. It felt like a thoroughbred to fly. The landing was then a gentle touchdown at 98mph and a rollout to a standstill in about 800yds.

In Petter's office the atmosphere was relieved but still professional. The pilot's report was listened to in silence, until he observed that the rudder hinge moment nonlinearity felt like over-balance. Petter agreed it could well be, but no one should jump to conclusions. By the end of 13 May, the decision had been taken to investigate the rudder over-balance problem by reducing the horn balance in stages. With the first reduction complete, the aircraft was flown on 18 May and it successfully responded up to 489mph at 15,000ft. Its control and stability was assessed as excellent. But another problem occurred at speeds in excess of 460mph: a vibration of about eight cycles per second.

After several more flights, on 26 May, using vibrograph sensors, an oscillation at 8cps was measured with some 24cps identified as elevator flutter. This meant increasing amplitude if the speed was not dropped immediately, resulting in diverging oscillations, which could destroy the aircraft. Beamont then took up Dai Ellis to confirm his suspicions, and, after holding the flutter speed long enough to impress his passenger, Beamont asked, 'What did you think of it, Dai?' 'Christ,' said the voice from the back, and added that the best place to investigate this was on the ground. The prototype was laid up until 5 July whilst modifications were made to the elevator horns and mass balances, and a final 'production' shape incorporated for the new rudder horn.

Between 6 July and 31 August the Canberra made thirty-six flights and handling clearance was successfully achieved over the initial design flight envelope. There was some small 'snaking' described, which turned out to be due to a turbulent wake behind the canopy, and was easily cured by adding a simple fairing. An altitude of 42,000ft was exceeded at Mach 0.8. The initial design speed of 540mph at 4,500ft was achieved, giving a margin of 23mph over the proposed initial service limit of 518mph. In all tests the smooth response and stable handling qualities were apparent. The low wing loading and the high power/weight ratio gave astonishing manoeuvrability at above 40,000ft.

The internationally renowned Farnborough Air Show was due to start on 6 September, and Petter (and Beamont) were both determined to display the Canberra to its full potential. So on 22 August, on the way back from high-Mach-number handling at 40,000ft, Beamont flew full rolls, rolls-off loops and full loops. He found the behaviour smooth and straightforward, though needing some muscle power at speeds above 400mph.

The aircraft was flown to Farnborough on 4 September, resplendent in a renewed and highly polished cerulean blue finish. On Monday the sixth the Canberra taxied out for the first demonstration, and with all eyes upon it the port engine died. The unburnt fuel flowing through the hot tail pipe caused an enormous amount of grey smoke. Petter was appalled at this apparently dramatic technical fault, leapt over the crowd barrier, and ran across the runway to his ailing prototype. He ordered for 'all tanks to be filled up', but fortunately the design office engineer in charge (John Sarginson), took the responsibility of ignoring Petter's order. Subsequently Petter was unusually magnanimous in writing to Sarginson, saying 'you kept your head when others failed ... take three days leave'. At the end of the day the last item was the rescheduled Canberra launch, and it is best described by the pilot himself.[8]

This time with no problems 799 rolled into its take-off run, leapt into the air after less than 700yds, was held down until reaching 200kts, and then entered a 45° banked climb to the left, reversing this to the right and in a dumb bell turn through 800ft over the crowd, the blue jet bomber was brought back down the runway at 100ft at full power, then pulled up vertically when passing the tower at 400kts into a half loop followed by a 45° downward roll through 220° to dive with power off back to the western boundary, and then a left turn for a 360° roll along the runway from 100ft to 500ft. Then left into a vertical bank around the north boundary pulling a tight turn at 4G in front of the crowd before rolling out to the east and pulling up into a wing over, dropping to 150kts, and lowering the undercarriage whilst turning left handed over the crowd to come back in a slow left turn to line up with the runway just short of the tower at slow speed, rocking the wings. Throttle to full power, undercarriage up, then dumbbells wing over to the west to come back for the final item … pulling up from 380kts at 100ft in front of the tower to roll out at the top of a half loop, lowering the undercarriage and bomb doors for a high rate of descent left spiral down to land.

The aircraft suffered a malfunction when instrumentation boxes from the bomb bay disappeared, leaving just a handful of wires trailing from the fuselage.

This demonstration was repeated daily for the rest of the week, and it was obvious that this new aircraft had created a stir in the aviation world. In *Aviation Week*, the leading North American journal, the editor, Robert Holz, summed up the occasion:

The biggest military surprise of the show was the English Electric Company's sky-blue Canberra bomber. US observers were not impressed with the Canberra's straight wing, a somewhat conventional configuration on the ground. But in the air the combination of test pilot R.P. Beamont, and the 15,000lb thrust from two axial Avons, made the Canberra behave in a spectacular fashion. Its speed range from 500mph to less than 100mph was ably demonstrated by Beamont, who followed his high-speed passes on the deck with an approach using full flaps, gear down and bomb bay doors open, that slowed the Canberra to less than 100mph. At this speed he rocked the big bomber violently with the ailerons to show the full control available as it approached stalling speed. Beamont whipped the bomber, designed to carry a 10,000lb bomb load, around the deck like a fighter, flying it through a series of slow rolls, high speed turns, and remarkable rates of climb. The Canberra was originally designed for radar bombing at around 50,000ft but Beamont's demonstration convinced many Britishers that the new bomber may prove its versatility at everything from low-level attack through to night flying to high altitude bombing.[8]

This view was not lost on the USAAF authorities, who later chose to have it built in the US against competition from several American companies.

The *Aviation Week* article was written some time after all four A1 prototypes, including 799, had flown; it was only later in 1949 that the name Canberra was chosen, by George Nelson. The traditional way of naming bombers presented a problem. All coastal town names were reserved for the Navy. River names had been reserved by Rolls-Royce. Beamont had said 'we can't call it the Chorlton-cum-Hardy'. George Nelson had noted the Australian interest in the A1 so chose Canberra. (It was later, in 1951, that the bomber was christened by Sir Robert Menzies.)

Before the Canberra was handed over to the RAF, there were two problems to be tackled. First of these was the runaway of the tailplane incidence angle following activation by the pilot. Since the powered tailplane incidence was much more powerful than the pilot/elevator combination, a potentially fatal pitch-up or pitch-down manoeuvre would result. It was thought that the runaway was due to sticking of the electric actuator switch following a trimming operation. It was fixed with a duplicated trim switch and a redesign of the wiring and actuator stops. Strangely the problem persisted in the conversion of Canberra to Martin B-57 in the US, where it caused fatal crashes.

Another problem in early development flights was the 'freezing' of the ailerons at high altitude. The first time that Beamont experienced this problem he nearly had to eject, but as he descended the ailerons freed themselves with the help of some vigorous wheel movements. This cause of problem was identified as the nose balance using a very close gap behind the shroud, rather than a diaphragm as in the Westland–Irving balance. At altitude the differential contractions due to temperature changes caused the small gap to close. This could be fixed by grinding the aileron lips where scratch marks showed on the shroud behind the rear spar. On production aircraft still closer tolerances were imposed for aileron assembly.

In conclusion it can be said that the Canberra design and development programme went very smoothly and the aircraft entered service with a good record of safety, reliability, and effectiveness.

Bomb aimer in nose.

During 1948 the relationship between Petter and Arthur Sheffield deteriorated significantly as the three versions of A1 were developed in the main works at Strand Road. 'Sheff' had the problem of converting his factories from being involved in wartime mass production to ones creating a miscellany of products ranging from experimental aircraft to diesel engines and electric cookers. The production aircraft for the new specifications B5/47 and PR31/46 differed significantly from the prototypes to specification B3/45. Because the high-altitude radar equipment was going to be too big for the Canberra, the aircraft had to resort to visual bomb aiming by an additional crew member in the fuselage nose.

In other words a wholesale redesign of the front fuselage. In 1949 the pressure became intense. All four prototypes began their flight development programmes, and production orders were received for 130 B2, PR3 and T4 variants. Towards the end of 1949 the relationship between Petter and Sheffield had become impossible, and Petter forced the issue of who should build experimental aircraft to Sir George Nelson, who did not see his way to giving Petter everything he desired: Petter threatened to resign. Frederick Page tried to persuade him to stay on, with a promise to join Petter in fighting the Sheffield establishment. He did not succeed. At the end of 1949

Petter only appeared once again at Warton, to speak to a few people, clear his office and tidy up some administration matters. He formally terminated his contract in February 1950.

Canberras in Operation

Immediately Petter resigned, Frederick Page was appointed as chief engineer, at the age of 33. The conflicts with Sheffield continued, who was told that Page was younger than Petter and would not give up as easily. In fact it took another twelve years, when English Electric was part of British Aircraft Corporation, for Page to take control of everything at Warton, Samlesbury and Preston. At this point Sheffield was excluded from all aircraft work.

Although Petter resigned in 1950, and did not participate in the subsequent developments of the three versions of the Canberra, these aircraft were very much his babies. Therefore it is worth describing their progress, after his departure, into the company's most successful operational aircraft, although Frederick Page must be given the credit for leading the team which saw this through.

The B5/47 was the bomber version, and between May 1951 and September 1952 English Electric delivered a total of 206 Mk II bombers to the RAF, and a further 210 were subcontracted to Avro, Handley Page and Short Brothers. This made a total of 416 delivered to Bomber Command, making this model by far the most prolific of the Canberra types. The RAF conducted exercises with the Canberra using Meteors and Venoms as enemy fighter threats. The Canberra had an 8,000ft altitude advantage and so was invulnerable to these fighters and others of similar performance in the Soviet Union. However, bomb delivery at high altitude was inaccurate using visual bomb aiming. For example in a night attack on Egyptian airfields during the Suez war in 1956, only three airfields out of seven were put out of action. These Mk II bombers were eventually replaced by V bombers in 1957.

In 1952 a Canberra Mk V was developed for the role of a low-level interdictor. It eventually became the primary operator for a bomber in the Tactical Air Force. It was armed with various combinations of guns, rockets and bombs in the bomb bay and under the wings, and had wing-tip jettison fuel tanks. This was a major departure from the original design specification, and looked more like Petter's original concept for a twin-jet fighter–bomber. In fact the pilot's seat was raised inside a fighter-type canopy, offset to the left to improve visibility.

Offset 'fighter' canopy to improve visibility.

The versatility and manoeuvrability of the Canberra was exploited when the Tactical Air Force was employed to carry US nuclear bombs. These had a large yield so an extraordinary manoeuvre was required to deliver bombs from a low altitude, known as the Low Altitude Bombing System (LABS). The aircraft flew a low-level approach at high speed, followed by a sharp pull-up into a loop, during which the bomb was released. The Canberra would then roll out at the top of the loop and dive at high speed to a low level sufficiently far away from the detonation to survive the blast and thermal shock. Such a manoeuvre made this aircraft have a greater chance of survival than other vehicles of this era. A total of 199 interdictor Canberras were delivered from English Electric and Short Bros.

The Canberra PR3 was the photo-reconnaissance version, able to fly above the maximum altitude of enemy fighters, collect intelligence data necessary for strategic planning and strike operations. It could also play a part in such strike operations by dropping target illumination from a wide range of altitudes. The modifications for this role were relatively straightforward: a longer fuselage to accommodate extra fuel, a camera bay and a flare kit. It turned out that this new fuselage was so much more flexible as to induce serious resonant vibrations. It was therefore stiffened by a series of external surface plates bolted to the rear fuselage. The RAF took delivery of thirty-five PR3s in the early 1950s to replace a squadron of Convair B-36D aircraft. This was a larger vehicle with a range of 8,000 miles but an altitude limit of 43,000ft, which had become within the range of the current Soviet interceptors. The PR3 was followed by the PR7, which had Rolls-Royce Avons with a thrust of 7,500lb compared with the 6,300lb in the PR3. Integral fuel tanks in the wings gave a range increase from 3,600 to 4,300 miles. English Electric delivered seventy-four of the PR7s by the end of 1954.

In 1953 a contract for a PR9 version was awarded, for an operational ceiling above 60,000ft using Avons of 11,250lb thrust. The wing area was increased by 10 per cent, mostly by increasing the chord between nacelles and fuselage, and partly by increasing the span of the outboard wing. The first flight tests showed a considerable improvement in the rate of climb but the effective ceiling was only 55,000ft. This limitation was due to the increased drag of the new wing. Wind tunnel tests showed that this was due primarily to the discontinuity between the inner and outer wing. The critical Mach number was now surprisingly reached in the outer wings. This aircraft was therefore the only one of the entire Canberra family which did not meet its specification. It also had a structural failure when demonstrating a 5G pull-out from a dive at maximum speed. The increased bending moment in the longer wing had caused a failure in the wing–fuselage attachments. These were redesigned and showed to be safe by Beamont performing a similar manoeuvre in January 1960. Subsequently twenty-two PR9s were delivered to the RAF and were used for nearly twenty years, until the early 1970s.

The T2/49s, the third of Petter's inspired derivatives, were the trainer versions, to provide pilot conversion training for the Canberra family, since many of the new pilots had not previously flown jet aircraft. Dual control was provided for two pilots side by side inside a broader double-bubble canopy. There was only just enough room and it was cramped. The navigator's station behind the pilots was retained. Altogether sixty-eight trainers were delivered, most going to the Operational Conversion Unit at Bassingbourne.

Finally mention must be made of the astonishing number of world record flights captured by the Canberra.[8] During its first ten years the Canberra was hardly ever out of the newspapers as a record breaker. By 1958 it had achieved no fewer than nineteen time and distance records and three altitude records. Both the bomber version and the photo-reconnaissance version took part in these records. Two altitude records of 66,000ft were captured using Bristol Olympus engines, and a record 70,300ft was achieved with the aid of a Scorpion rocket motor. The use of the Canberra as a jet- and rocket-test bed at high altitude proved the remarkable handling and control of the Canberra in conditions far beyond those for which it was designed.

The Martin B-57[8,13]

The biggest compliment the Canberra was to get was the decision by the USAAF to buy 400 aircraft in preference to several American fighter–bombers competing for the same specification. This is a story worth telling, not just to show the aircraft's special qualities, but also to illustrate the major impact made by its test pilot, Roly Beamont.

In August 1950 a message was received at Warton Flight Operations that an important demonstration was required for some American VIPs. It was to take place at RAF Burtonwood, and Warton was asked to comment on the practicability of this. This was a time of total concentration in the test programme team as they had a target time for entry in to the RAF service of spring 1951. Warton responded that a demonstration was possible but a support ground body could not be provided, so a pilot could land, take off, and demonstrate and then fly back to Warton, all without stopping engines. The Americans were determined to see this aircraft so, after some planned test flights on the morning of 17 August, Beamont flew VX169 the short trip to Burtonwood, in poor weather conditions of drizzle and a leaden sky. With no traffic on this almost deserted post-war airfield, VX169 was landed and taxied over to a small group of raincoat-clad people with staff cars. One by one this group of earnest civilians and uniformed USAAF generals peered into the cockpit.

Air Commodore Strang Graham, the English Electric sales manager, gave a thumbs up signal for the demonstration to start. It had stopped raining, and with full power, and less than 3,000lb of fuel, the Canberra was airborne in less than 700yds. It then pulled up steeply into its characteristic 'cartwheel' wing over to come back down the runway and pull up at 300kts into a roll-off, and then begin a display uninhibited by traffic. Some minutes later, with the fuel gauges showing a need for early landing, VX169 gave the attentive group a final low pass and headed north for Warton.

The Warton test team had almost forgotten this episode, as they were absorbed in their own test programme, but in January 1951 the Americans, despite advice from their government that the Canberra was in too early a stage in its development, insisted on a full evaluation in competition with their own emerging jet-bomber prototypes. These were the:

Martin XB-51
North American B-45
Propeller-driven North American AJ-1

Douglas A-26

Canadian Avro CF 100.

The competition was for an aircraft to meet a USAAF requirement for a medium-range 'Intruder'.

The RAF agreed to release one of their pre-production aircraft, WD932, and to deliver to Washington in February. Squadron Leader Arthur Callard and two crews delivered WD932 to Andrews Air Force Base Washington on 21 February, after completing the first direct jet crossing of the Atlantic in a record-breaking time of 4 hours 37 minutes from Aldergrove to Gander. Beamont took the aircraft over to Andrews the following day.

The briefing for the competitive display for the Pentagon board took place on 26 February and was strictly formal and 'by the book'. Beamont (the only non-American pilot present) was concerned that each was to fly the same set pattern of simple manoeuvres in a 10-minute time slot, and he eventually asked if the pattern could be varied to suit particular aircraft. 'No,' he was told firmly. 'This is an Air Force trial and not a Farnborough show.'[8] He then realised that the Canberra could complete the set manoeuvres in about half the time, and no one had said anything about what to do with any time saved, so he resolved not to waste it.

The American aircraft all flew their routines safely but unimpressively, owing to their high wing loading and modest thrust-to-weight ratios. Only the XB-51 looked new and impressive but its high wing loading and low manoeuvrability limited its presentation severely. The Canberra was the last to fly and WD932 flew through all the set manoeuvres; when it was completing a final tight turn to approach with gear and flaps down, it had 4½ minutes left of its 10-minute slot. So, raising the undercarriage and flaps, and opening up the fast-accelerating RA3 Avons, Beamont zoomed over the spectators, pulled into a minimum radius 360°-turn directly overhead, half rolled down into a 400kt fly-past and backed up into its 'cartwheel' wing over. He followed this with a power-off-with-airbrakes tight spiral dive to pull out in front of the spectators and then went up and over, whistling almost silently, before he lowered landing gear and flaps once more, went in to a tight left bank and landed. A voice from the navigator's seat of the USAAF evaluation team said, 'Gee, that's more manoeuvrability than I've ever seen.'[8]

The XB-51 was about 100mph faster in a straight line than the Canberra, but that was it. There was no contest in overall ability to meet all the required criteria, and the Canberra was declared the winner.

The Pentagon was then faced with a decision about its manufacturer. The Glenn L. Martin company did not have a very full order book, so was selected. The contract was to build 250 Canberras, given the USAAF-type designation B-57A Canberra, and the licence agreement was drawn up by May 1951. In June 1951 the USAAF needed two Douglas DC-4s filled with Canberra drawings to be taken to Martins for conversion to US measurements. A key figure in organising this collaboration between English Electric and Martins was Glen Hobday, who had been with Petter at Westland and joined him at Warton TC in October 1945. He was quoted as saying, 'I would be working for a man I really respected both technically and as a boss.'

The first production aircraft had its maiden flight in July 1953, a mere twenty-eight months after the contract had been received. Beamont flew this aircraft a little later and declared that it handled exactly the same as the B2 produced in the UK. The Canberra went on its way to equipping the USAAF, in which it still served thirty years later.

Of the 400 B-57s, most served overseas, in many roles, in countries such as Korea, Vietnam, and Pakistan, but in 1966 General Dynamics rebuilt several to fly at altitudes greater than 70,000ft. The wing area was trebled, and the engines installed were Pratt and Witney turbofans, supplemented by two smaller turbojets to be activated only at the higher altitude. Eventually the Lockheed U2 became the supreme high-altitude reconnaissance aircraft.

Some 901 English Electric Canberras were built and used by the RAF, and air forces of India, Venezuela, New Zealand, Rhodesia, Argentina and Peru. Some forty-eight were built in Australia for the Royal Australian Air Force.

The Canberra was retired by Bomber Command in September 1961 but the German, Cypriate and Singapore squadrons continued in their nuclear role. The Singapore-based squadron was the last to disband, in 1970. The photo-reconnaissance versions were equipped with Long Range Optical Photographic Cameras, enabling photographs to be taken deep into Eastern Europe whilst the aircraft flew along the German border.[15] Reconnaissance also played a key role in military operations in Bosnia, Kosovo, Iraq and Afghanistan.

Petter would have had mixed feelings about his beloved aircraft being used as a weapon of mass destruction, and even more so when used by the Argentines against the British in the Falkland Islands.

5

THE ENGLISH ELECTRIC YEARS: THE P1 AND LIGHTNING, 1948–1950

In 1946 the Atlee government decreed that powered supersonic flight was too dangerous and cancelled the Miles 52 research aircraft, which at the time was well advanced and would have given British aviation a substantial lead in supersonic flight, or at least defined the performance 'envelope' approaching it.[18] The Miles aircraft in some ways bore a resemblance to the American Bell X-1 which was to become the first piloted aircraft to exceed the speed of sound in 1947. The M52 was to have been powered by a 17,000hp Power Jets three-stage gas turbine and it was envisaged that it could climb to 36,000ft in 90 seconds, and attain supersonic flight at heights in excess of 50,000ft. The pilot was housed in a pressurised nose cone attached to the fuselage in the centre of an annular intake ducting. The whole nose cone could be ejected by explosive charges in the event of an emergency. The short span laminar-flow wings were extremely thin and unswept.

In 1946 Petter made provisional sketches of a supersonic fighter, keeping the design simple and strong. It had a small frontal area, achieved by mounting two Rolls-Royce Derwent engines one above the other in the fuselage. The prone piloting position was not favoured by Petter or his chief test pilot and in any case presented little advantage with this two-engine configuration.

Early in 1948 the Canberra prototype's first flight was still a year away, and Petter's team was almost fully occupied with its manufacture and development. However, a certain civil servant, who was to become one of

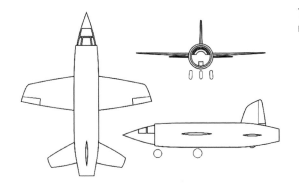

The proposed Miles M52 research supersonic aircraft.

the country's most influential aeronautical engineers (including becoming group technical director of the British Aircraft Corporation) was planning the prospects for a supersonic fighter. Handel Davies had just left the Royal Aircraft Establishment, where he had worked on the aerodynamics of the Spitfire, Mosquito and Lancaster, and became head of the establishment's Aerodynamics Flight Division. Later, in the mid 1960s, he steered Anglo-European projects such as Jaguar, Tornado and eventually Concorde.

However, his concern in 1948 was for the UK to produce the first steps for a supersonic fighter capable of Mach 2 in level flight. It was apparent that a precursor programme would be needed to identify a feasible configuration and design layout before a commitment to an operational programme. Handel Davies informed English Electric of an experimental research opportunity under specification ER 103, and enquired whether they had the staff available to join a group of aircraft companies who were looking at the potential design concepts for a Mach 2 aircraft. Petter responded with enthusiasm, and the very next day Handel Davies and his ministry team flew to Warton. They brought with them the ER 103 specification, which visualised a research and development programme to explore design options followed by ground tests and prototype flights in the first contractual phase. If successful, a follow-on contract would be placed for an operational interceptor capable of Mach 2 in level flight and able to climb rapidly to high altitudes. The restriction to Mach 2 was set since at this speed the stagnation temperature at an aircraft's leading surfaces, due to kinetic heating, is about 120°C, which is the limit for light aluminium alloy materials. Most supersonic aircraft for the next two decades (including Concorde) accepted this limit.

It was believed that for Mach 2 an operational aircraft would need two engines with reheat for take-off and acceleration, but should be capable of supersonic speed in level flight without reheat. The ER 103 specification called for Mach 1.2 to 1.5 in the first phase. The wing sweep was not defined and English Electric was invited to define its own version of the research and development programme to achieve this aim.

In the early summer of 1948 Freddie Page was set up to direct a team to lay out designs for an English Electric proposal, with Ray Creasey leading the critical area of aerodynamic design. It was apparent by then that the only feasible location for the engines would be in the fuselage, with ducts leading to an intake in the nose. This was similar to Petter's early design concept for the Canberra, but this time there was no conflict with a need for a fuselage bomb bay. An intake in the nose was envisaged because fuselage side intakes near the wing root might reduce the length of ducting needed but could suffer serious flow separations. It would be nearly a decade before boundary layer bypass designs became the solution to intake separation problems (such as in the Convair F-102 and later the Sukhoi Su-15, for example). A pitot intake would provide an adequate intake efficiency up to Mach 1.5 but above this speed the inlet shocks did not interact optimally, so for Mach 2 a body of revolution was needed with a centre body, the design of which was then understood.

The use of two engines, vertically stacked in a staggered layout, had the advantage that if one failed at take-off, there would be no asymmetrical thrust. There was the disadvantage that little volume would be available in the fuselage for equipment and fuel. This would not be a problem for the experimental research aircraft, and this configuration, designated P1, was intended in July 1948 to satisfy the requirement for a Mach number of 1.5 using a leading edge sweep of 40° for the main wing, and a similar sweep for a tailplane mounted on top of the fin.

Early on it was known that a highly swept leading edge would start to shed a vortex at a large enough angle of incidence. As the incidence increased so did the strength and extent of the vortex. The overall effect was that the lift continued to increase nonlinearly with incidence rather than suddenly separated, as in an unswept wing. Thus there was no classical stall, a possible advantage, but would large angles of incidence shield the fin and tailplane? A rather bizarre option at this stage was the choice of twin highly swept tailplanes away from the fuselage wake, each tailplane having equally swept fins. This would have been a structural nightmare to build and the design was never pressed after first thoughts were submitted to the

Schematic of the
English Electric P1 in
July 1948.

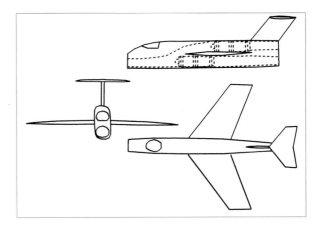

Ministry of Supply in October 1948. Petter wrote to H.M. Garner (director of scientific research at the ministry) requesting a meeting for 17 December and asking for a further contract to continue with the design work and a flying prototype.

A configuration for high-speed subsonic aircraft which was becoming popular at this time was the tail-less delta; examples include the Fairey FD2 and the Vulcan bomber in Britain, the Mirage in France, and the transonic Convair F-102 in the US. These aircraft achieved sufficient longitudinal stability and control from their 'elevons', and so did not need a horizontal tailplane. However, it was well known that the aerodynamic centre would move aft as the aircraft went from subsonic to supersonic, and there was doubt whether elevons would be able to retain stability and provide enough control power for the manoeuvres required for an interceptor at altitude. The P1 design was therefore changed to have wings with a sweep angle of 60°, both main wings and tailplane. The choice of 60° was straightforward because at Mach 2 the flow normal to the leading edge would be just sonic (see Appendix 1). However, at the wing–fuselage junction the isobars would not be at this angle of sweep, but would be unswept. There are two solutions to this problem. One would be to alter the wing thickness or camber at this point, the second would be to increase the local sweep angle for a wedge at the leading edge. This latter was chosen. (Later on in the development period the decision was made to change to the former.) The main wing was placed low in the fuselage and the tailplane was placed lower down the vertical fin. This was the favoured configuration in February 1949 for the P1 proposal.

Early model with highly placed twin fins and tail planes.

A very thin wing section would be necessary to reduce the drag enough to reach Mach 2 in level flight, so a thickness/chord ratio of 5 per cent was selected. This left enough depth for structural stiffness and provided enough space for the undercarriage to retract. The efficiency of conventional ailerons at such a high sweep angle was doubtful, so instead ailerons were placed at the tip of the wing normal to the flight direction, a strategy unique to this aircraft.

English Electric and Fairey Aviation submitted proposals, the former designated P1 and the latter FD2. The Fairey design was purely a research aircraft, although later it did go on to capture the world's air speed record. On 29 March 1949, the Ministry of Supply wrote to Petter with notification that preliminary approval was granted to proceed with design work on the P1. The authorisation was as follows:[19]

I confirm that we wish you to go ahead with the design work you propose, namely:

1. Manufacture and test to 1/8 scale a model in your 9ft × 7ft wind tunnel.
2. Make and supply to Rolls Royce a 1/4 scale model for testing the ducting.
3. Make and test a 4 inch model for high Mach number tests in your own high speed tunnel.

We are asking the Director of Contracts to cover you at £10,000 to ensure no delay for any part of this programme. Construction of prototypes in six months.

Schematic of the P1
proposal in February
1949.

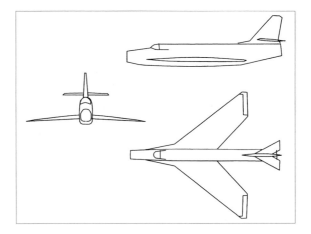

Petter replied on 4 April:

> I am glad to say the 1/8th scale model is finished and will be going in to the
> wind tunnel any day now. We are studying carefully the question of sweep
> back [the Royal Aircraft Establishment had advised not to exceed 45°] and
> think there may be good reasons for sticking to the present proposal, subject
> of course to the wind tunnel tests.

It is worth noting here that Sir George Nelson had, in agreeing with
Petter, decided that English Electric needed a selection of wind tunnels
better than any available in the Royal Aircraft Establishment. This foresight
now paid off. Dai Ellis and Roy Fowler had designed and made a 9ft × 7ft
tunnel for testing the jet bomber. A small water tunnel was made that could
produce high Reynolds numbers (see Appendix 1), which was desirable in
a small wind tunnel aiming to simulate full-scale aerodynamics, especially
for possible separated flow at high incidence or flaps down. It also allowed
easy observations of flow patterns by injecting aluminium particles into the
flow. Finally a high-speed tunnel was built using an old Rolls-Royce Nene
jet engine with a long diffuser. To confirm the predicted performance,
Petter approved funds to perform some small-scale testing of wind tunnel
configurations using compressed air. The results were sufficiently promising
to get the company funding to go ahead with the full-scale design. Using
the sheet metal fabrication resources at the Strand Road works, the tunnel
was completed in a few months. It had a working section of 4ft by 1ft and

was originally used for the Canberra at a speed up to M = 0.8. For the P1 the tunnel was fitted with a new diffuser and a working section reduced to 1ft by 1ft, incorporating slotted walls to enable testing up to M = 1.2, a unique feature at the time.

The noise of the Nene operating at ground level was a major problem, so the tunnel was set up at the far side of the Warton airfield, and the jet exhaust was pointed out across the Ribble estuary. Rolls-Royce was very supportive of this novel concept, but insisted that one of their test bed operators should act as a 'tunnel driver' to prevent the engineers from blowing the engine and themselves to pieces. One day Roy Fowler received an urgent call to go to Petter's office and found a local farmer, who was claiming that his cows in the field downstream of the tunnel were being upset by 'the abominable screech of your God-damned engine', and he was sure their milk production would be affected.[3] Surprisingly Petter expressed concern and would consider compensation if the noise could not be abated. He then ordered Fowler to do something about it. However, when the tunnel was run the next day, Fowler happened to look out and was surprised to see the cows converging into the jet efflux and congregating as close as possible to enjoy the warmth. Fowler took photographs and never heard from the farmer again. Teddy never asked what had been done to solve the problem.

A project design group was then set up under Freddie Page to undertake preliminary design work with engineers on temporary assignment from the Canberra design teams. This organisation continued for about four months, by which time the structural and control concepts had reached the stage for detail designs to begin. The formal Ministry of Supply contract was received on 12 May 1949, which happened to be the day before the Canberra's first flight.

Meanwhile the P1 layout was being tested in the low-speed wind tunnel, and at this stage had a low main wing passing through the fuselage in front of the lower engine and a tailplane mounted low in the fuselage, below both engines. The wing plan form had been settled, the spar positions decided and the main undercarriage wheels shown to be retracted inwards and housed in the wing root structure. The 75° swept root leading edge wedge had been deleted because it increased the pitch-up tendency. This configuration was used in the brochure submission dated 1 October 1949.

Petter wrote to the principal director of scientific research at the Ministry of Supply, sending sixteen copies of this up-to-date proposal, stating that:

extensive design, wind tunnel tests, a mock up, plus considerable work on structural problems, have been carried out under the design study contract. It is felt that the stage has now been reached at which it would be possible usefully to proceed to an advisory design conference and thereafter to the design and construction of a prototype aircraft.

It was about this time that a feature became apparent in the low-speed wind tunnel tests. A powerful vortex in the air flow separated over the upper wing surface near the leading edge at about halfway along the wing. This started at an incidence of a few degrees and then developed with increasing strength, moving inboard as the incidence was increased up to 30°. The strong rotation of the vortex flow was not detrimental to the wing lift or drag but it did create a large downwash in the tailplane region that made the tailplane ineffective in achieving longitudinal stability; in fact, the stability was better without the tailplane than with it.

Ray Creasey's immediate concern was that this vortex could be caused artificially by laminar boundary separation due to the low Reynolds number of the wind tunnel, and so may not occur in the full-scale aircraft tests. Attempts to fix this boundary layer transition with trips along the wing leading edge made no difference to the behaviour of the vortex. This fact gave the aerodynamicists some encouragement for faith in their tunnel tests.

Moving the tailplane to the top of the fin made the stability even worse, but a location below the main wing plane provided good stability

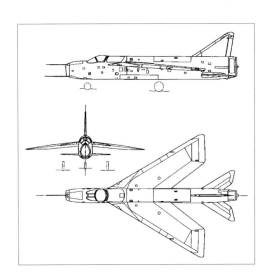

General Assembly assignment of P1 in late 1949.

and remarkably linear aerodynamics over the required range of angles of incidence. An unswept wing behaves linearly as incidence is increased, until it reaches a stalling angle, when separation and a sudden loss of lift occurs. For wings with a highly swept leading edge, the vortex-generated lift does not behave in this fashion. It is already separated, and the vortex strength and lift continues to increase (nonlinearly) with incidence up to values of 30° or more. The only downside may be the pilot's vision when landing or taking off at such a high incidence. The most famous example became the civil aircraft Concorde, which adopted a mechanically actuated 'droop snoot' to give the pilot a decent view when landing. The P1 cockpit was deliberately close to the aircraft's sloping nose.

The low position of the tailplane was now accepted but the main wing was moved up by about 22in to increase the vertical separation of the wake from the tailplane. This meant a rearrangement of the two engines and their ducts. The drawing office had to start all over again, but at this stage no metal had been cut. It was even possible to tilt the fuselage nose down by 2° relative to the wing to improve the pilot's view at these high incidences during the landing approach. A leading edge flap, capable of drooping, was incorporated at the wing root, and later a notch would be added to the wing leading edge at about two-thirds of the semi span to stabilise the point of vortex separation at the higher Reynolds number in flight, and to improve stability at high angles of incidence. (During test flights Beamont showed this flap was unnecessary and it was removed.) This then was the definitive-version P1A, which was hardly altered and flew five years later as the WG760 with immediate success.

With such a swept wing design, any side slip or yaw causes higher lift on one wing than the other. The wind tunnel tests showed that the consequent rolling moment would be unacceptable, and could only be cured by a large increase in the wing anhedral. However, when the technicians checked their figures, they realised that a mistake had been made when calibrating the wind tunnel balance. This error was not discovered for about a month, during which time the wing root structure and undercarriage were being redesigned. The tunnel operator, Roy Fowler, realised his responsibility and confronted Teddy Petter. He was always an attentive listener, and, after this confession, the expected blast did not occur, but Petter said, 'Well, Fowler, I'm glad you've found this out and brought it to my immediate attention. Now you can come and help me sort out the mess you've made in the design office.' Later Fowler said that this was typical of the man.[3] He was

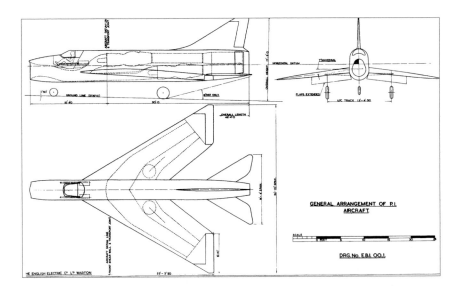

G.A. of P1A built later in 1954, as WG760 and WG 763.

quick to forgive honest mistakes if immediately admitted, but terrible in his anger if he felt you were not being forthright or were attempting to transfer the blame to others.

Although the ministry appeared willing to accept the latest design, the Royal Aircraft Establishment was not. A sweep of 60° was considered too dangerous, particularly at landing and take-off. They favoured a 50° sweep and a high tailplane location out of the downwash. They then persuaded the ministry to contract with Short Bros to construct a low-speed research aircraft, the SB5 (later WG768), in which the effects of sweep and tailplane location could be studied in flight.

Three different sweep angles could be selected (but not in flight): 50°, 60° and 69°. High and low tailplane positions were possible, and a single Derwent engine was used in this 7/8 scale aircraft, which made it seriously underpowered: it took twenty minutes to attain 7,000ft. By the time that the WG768 first flew, in December 1952, the P1 prototype was taking shape and could not have been altered. The SB5 with 50° and high tailplane was a disaster, the pilot could not even slow down to stall because of serious buffeting due to turbulent vortices. The wing was then moved to 60°, which removed the buffeting but pitching stability was poor with the high tailplane. Not until 1953 was it rigged up with the P1 sweep and low tailplane, whence it became almost a flying machine. The whole exercise was a complete waste of taxpayers' money, although it did give Beamont some valuable flight handling experience at low speeds before he came to fly the P1 prototype.

The Ministry of Supply now had sufficient confidence in the outcome of the P1 programme to issue an operational concept specification, designated F23/49. The need for extra thrust using reheat (in which neat fuel was ignited in the exhaust ducts) was accepted; and the consequent short duration of a mission. It specified that interception would be made using ground control, and the aircraft radar could lock on to a target. Ground control would then tell the pilot when to launch its Red Tops missile. English Electric consequently redesigned the nose of the P1 to meet this requirement by installing the radar in a cone-shaped centre body.

By now the detailed design of the P1 was well underway, and it was a sea change compared to the basic simplicity of the Canberra. The wing structure was a good example of the novel solutions. Usually the torsion box had stiffeners in the wing covers to raise the buckling stress, but the P1 wing was so thin that such stiffeners in the top skin would meet those in the lower. The wing therefore became a multi-spar construction, but with several discontinuities to accept the retracting undercarriage. The front spar was also curved at the fuselage to make way for the short leading edge flap. The design and analysis of such a complex and highly loaded structure was a remarkable achievement for the stress office at the time. It was no surprise that English Electric designed, built and used the new digital computers for structural analysis; and later its structural engineers designed the first finite element code (now obligatory for all military and civil aircraft).

The powered controls, including an all-moving tailplane without elevator, were all innovative, and were to require much testing and development in the years ahead.

Short Bros research aircraft to study effect of sweep and tail plane location.

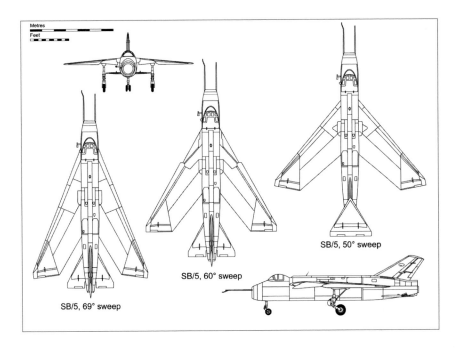

Schematic of SB5 with three different sweep angles. (Emoscopes)

In February 1950, before the various redesigns were complete, Petter resigned his position as chief design engineer. He would not be able to claim that the P1 was his 'brainchild' in the same way that the Canberra was, but he did oversee the first critical stages of the development of the first British jet aircraft to fly supersonically and achieve Mach 2 in level flight. Without his contributions to English Electric's reputation they would probably not have been given the opportunity to participate in what became the P1B and Lightning programme. He showed great faith in his technical colleagues and in particular did not question their judgement and recommendations regarding the complex and innovative aerodynamics and the design layout. Fortunately later events proved his faith to be justified. Freddie Page said, 'What a pity that he would not stay and fight with his Warton team. As with Camm, it was a privilege for me to have been a member of his team.'[3] His chief test pilot, Roland Beamont, went further:

> Teddy Petter's brilliance as an innovative engineer and designer has been
> much under-recognised and his 'controversial' character over-emphasised.

He was certainly a stubborn and over-bearing man and strongly confident in himself, but he was able to recognise strong views in others if they were sound, and then support them. English Electric lost a great technical leader when he left. Those who served him will always remember him as the vital element in Britain's first jet bomber and Britain's first fully supersonic fighter.

The P1, P1A, P1B and Lightning

In April 1950 English Electric received a contract for the construction of three P1 airframes, under the leadership of Freddie Page as chief engineer and Ray Creasey as his deputy. It is therefore appropriate to see what other countries were designing and compare with the P1A (and later the Lightning, which did not differ very much from the P1A).

Contemporary design alternatives:

- **The P1A** First flew in 1954. Top speed Mach 2.1, range 850 miles, service ceiling 54,000ft (later 85,000ft in the tropics) rate of climb 50,000fpm (with Avon engines) thrust/weight ratio 0.78, wing loading 76lb/sq. ft.

- **The Lockheed 104 Starfighter** First flew in 1954. Top speed Mach 2.0, range 420 miles, ceiling 50,000ft, rate-of-climb 48,000fpm., thrust/weight ratio 0.54, wing loading 105lb./sq. ft. This was the first truly operational supersonic fighter, and over 2,500 were built, many under license by various foreign manufacturers. It achieved its performance by having a light structure and an exceptionally streamlined low-drag weak shock-wave fuselage. Its small ultra thin

Illustration of P1B showing complex wing structure. (*Flight*)

P1A first flew in 1954.

wings (3.36 per cent t/c ratio) carried an incredible high wing loading. It had consequently poor manoeuvrability and was difficult to fly. It needed blown flaps to land, and had the worst safety record of any modern NATO fighter aircraft.

- **Mikoyan-Gurevich MiG-21** First flew 1955. Top speed Mach 2.05, range 150 miles, ceiling 58,400ft., rate-of-climb 44,280fpm., thrust/weight ratio 0.47, wing loading77lb./sq. ft. This must have been the most commercially successful military aircraft ever. Over 11,000 were built and it was still in service in Russia until the 1990s! Like most contemporary designs it used a delta wing plan form but also a separate tailplane, so it was straightforward to fly. It had an overall impressive performance and high rate of climb, but had poor manoeuvrability and a short range.
- **Dassault Mirage III** First flew 1956. Top speed Mach 2.2, range 746 miles, ceiling 50,000ft, rate-of-climb 16,000fpm., thrust/weight ratio 0.6, wing loading 56lb./sq. ft. Dassault Aviation produced a versatile aircraft capable of ground attack, reconnaissance, and training. Over 1,400 were built in France and a number of other countries. It was the first European aircraft to exceed Mach 2 in level flight. It had no tailplane and consequently could not use conventional flaps, but relied on the high stalling incidence of a delta wing, and therefore it needed a long take-off run. Its rate of climb was poor so it was not a decent interceptor.
- **Sukhoi Su-15** First flew 1962, Top speed Mach 2.5, range 900 miles, ceiling 59,000ft, rate-of-climb 45,000fpm., thrust/weight ratio 0.48, wing loading 96lb./sq. ft. This was not strictly a contemporary design, since it did not fly until 1962. It was a two-engined interceptor, and

Lockheed 104 Starfighter.   MiG-21.

had learned the art of designing two side intakes with boundary layer escapements. Its rate of climb was high as its mission was to intercept the Boeing 52 bombers. Although it had tailplanes, it needed blown flaps to land, and was not easy to fly. Over 1,300 were built and it was the main fighter resource until the demise of the Soviet Union.

A comparison of the P1A/Lightning with the above contemporaries is shown in the table.

	Lockheed 104 Starfighter	MiG 21	Dassault Mirage III	Sukhoi SU-15	P1/Lightning
Wing loading (lb./sq. ft)	105	77	56	96	76
Thrust/ weight ratio	0.54	0.47	0.60	0.48	0.78

The superior thrust/weight ratio stands out and clearly would enable the aircraft to achieve a high rate of climb and intercept capability. The wing loading is not too high and should lead to reasonable manoeuvrability and landing/take-off performance. The rate of climb made all high–altitude Russian bombers vulnerable. The only competitor for speed and rate of climb was the Lockheed Starfighter, but its handling and manoeuvrability were not in the same class.

Mirage III.

Sukhoi Su-15.

The P1 was developed into the P1A; improvements included leading edge slots to control the vortex development, and small leading–edge flaps at the fuselage side. It first flew at Boscome Down in late 1954. It was a great success, and Beamont said 'this first flight had been a classic technical success, except for a complete breakdown in communications between the pilot, the chase aircraft, and air traffic control'.[17]

The next development stage was the P1B, the operational prototype, of which twenty were built. Developments included the new centre body housing the radar, and several aerodynamic refinements to improve low-speed and cruise performance, such as a wing camber and subtle change in the leading edge sweep in the outer wing to give a slightly larger wing tip area. A larger fin size was needed to improve directional stability at the higher Mach numbers. The first flight of the P1B took place at Warton on 4 April 1957. Ironically it was at this time that Duncan Sandys, the Minister of Defence, announced that no more manned fighter aircraft would be designed and built, and would be replaced in defence by missiles.

It would need a development period of more than six years from 1954 before the P1/Lightning achieved its original specification. One limitation, particularly if reheat was needed at interception, was the range, limited by too-rapid fuel consumption. A conformal ventral fuel tank was therefore added beneath the fuselage.

The production Lightning F1 entered service with the Air Fighting Development squadron in May 1960. Its performance lived up to its specification and pilots found it easy to fly. One strategy devised by RAF pilots was to accelerate to Mach 2 and then put the aircraft into a steep climb, culminating in a ballistic trajectory through the thin atmosphere, reaching heights in excess of 70,000ft. Eventually they succeeded in intercepting the U2, which the USAAF claimed to be invulnerable at such heights.

Initially its serviceability was poor due to the complexity of its systems. Further squadrons took delivery, and the 'next generation', F3, with Red Top missiles, flew in June 1962. More than three hundred entered service altogether.

During the 1960s the Lightning's range and firepower shortcomings became apparent. They were gradually phased out of service between 1974 and 1988, and were replaced by the Tornado (from the same stable) with a far larger weapons load and considerably more advanced avionics. According to the 'Lightning Association' in April 1985, Concorde was offered as a target to NATO fighters, including F-15 Eagles, F-16 falcons, F-14 Tomcats, Mirages and F104 Starfighters. Only a Lightning managed to overtake Concorde on a stern conversion intercept.

P1A taking off from Warton in 1954.

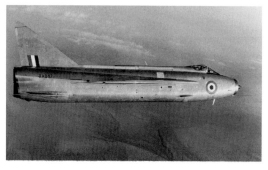

The P1B, Lightning, with wing camber, centre body intake, larger fin and dorsal streamlining behind cockpit.

6

THE FOLLAND YEARS: THE MIDGE AND THE GNAT, 1950–1959

The Midge

Petter resigned from English Electric in February 1950 and in October was appointed to the Board of Folland Aircraft as deputy managing director and technical director, and also managing director designate pending the retirement of Henry Folland due to ill health. It was effectively nine months since he had worked at Warton, and one can only speculate how he had been active in this time, apart from negotiating a deal with Folland Aircraft Ltd. He probably spent some time in his religious retreat, following his last turbulent months at Warton, but he was undoubtedly capitalising on his English Electric experience in forming a design for a new interceptor fighter, with two major lessons learnt after the design of the P1 and Lightning:

- A fighter designed to intercept high-altitude bombers from the Soviet Union had to have a thrust/weight ratio close to unity, and therefore the choice of engine was crucial.
- The Lightning could outperform any contemporary fighters but needed two engines and a total all-up-weight of 46,000lb to carry the necessary weapons and equipment. It was a complex aircraft and difficult for the RAF ground engineers to maintain and keep in operation. All this had made an impact on Petter's vision for a new fighter.

A third lesson was the realisation, after twenty years as a designer, that he could not work effectively under senior management, especially not being able to separate experimental manufacture from production assembly lines with all that control and quality assurance needed.

Folland Aircraft Ltd had been formed in 1937, when it had recruited Henry Folland, technical director of Gloster Aircraft Company, to rescue British Marine Aircraft Ltd, a company formed in 1936 to build large civil flying boats. New premises had been built at Hamble on the east bank of Southampton Water. Three large flying boats could be built simultaneously, and there was the biggest opening door in the world, giving access to a slipway running down to the water. Unfortunately the company had signed a deal with Sikorsky to build under licence the Sikorsky S42-A, which could carry thirty-two passengers over a range of 1,200 miles. However, the American parts did not come close to specification, and British Marine had to do a deal with Sir Ernest Petter. After this deal fell through, the board prepared to liquidate. However, Henry Folland and his new team turned the company around and by December 1937 the board moved to change its name to Folland Aircraft. From 1938 to the mid 1950s the company was financially sound, due to an extraordinary number of subcontracts for building components for the Beaufighter, Beaufort, Blenheim, Brabazon, Canberra, Dove, Hornet, Hunter, Mosquito, Seafire, Sea Vixen, Spitfire, Sunderland, Viking, Venom and Wellington.

By mid 1950 Henry Folland was planning to retire, leaving the production and financing to Tommy Gilbertson as general manager. Gilbertson was feared by all workers, who had to suffer the violence of his tongue if things did not go according to his plans. What Henry Folland needed was a chief designer of stature who could lead the company in the design and development of its own aircraft, and so provide some stability for its 2,500 workers when subcontract work fell off. This would also perpetuate the Folland name in the world of aircraft design. What Folland had to offer was capital and production facilities, and this was the legacy that Teddy Petter inherited. By 1951 his concept of a lightweight fighter was already well developed and the prospect of capital and production facilities meant that Petter, for the first time in his life, could proceed to design and manufacture without any financial support from the Ministry of Supply.

What Petter had in mind can be gleaned from his lecture to the Association Française des Ingénieurs et Techniciens de L'Aéronautique at Le Bourget two years later. Summarising his thesis, he argued that lightweight design was the key to large production numbers and the economy of scale. Capital, tooling and production costs of current fighters made it impossible to build large numbers. Tooling alone takes two years and the quantities are so small that the benefits of mass production are no longer achievable. He suggested a small fighter (5,500lb) compared to fighters like the North American Sabre (16,500lb). A small fighter can carry a payload of 1,000lb (seen as adequate) and can achieve a proper performance to intercept, provided a suitable engine is available. He compared the 5,750 man hours to build 1,000 small fighters with the 29,000 man hours for 200 large machines. A total cost of £6m gets 215 large machines or 900 small fighters.

He concluded that major cost reductions are possible for a fighting system to meet most (if not all) requirements for standard day fighters costing four or five times as much, and so many more NATO countries could afford them. He concluded: 'Great Britain is in danger of losing excellent pilots because it is unable to build a sufficient number of machines to use their skills and courage!'

The Folland works at Hamble, which Petter joined, included a beautiful Georgian mansion, where he set up his administrative staff and took over

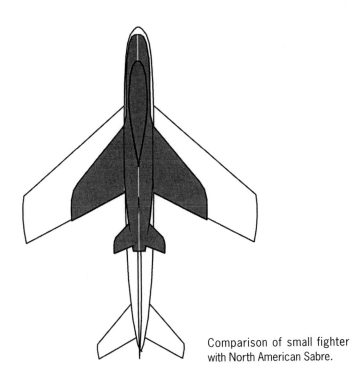

Comparison of small fighter with North American Sabre.

Henry Folland's impressive office (and conservatory). There was also an elegant dining room and wine cellar, and Petter became renowned for entertaining and carving joints or fowl in true baronial style, counter to his reputation for meanness.[3]

His immediate task upon joining was to recruit a design team to join Gilbertson and Gordon Hudson (chief of stress) who he had inherited. Gordon Hudson worked on the Gnat from the start, and later went on to become chief designer of the Hawker Siddeley Hawk. He said of Petter, 'Teddy was regarded as an austere and puritanical man, never given to the use of bad language; I heard him say "damn" once, almost under his breath and under extreme provocation. He was an outstanding designer, and although basically an aerodynamicist he was also an intuitive structural designer.'[21]

Petter had made a 'no poaching' agreement with Freddie Page but this did not prevent members of his English Electric team from resigning of their own free will. These recruits included Alan Constantine (designer), Roy Fowler (chief aerodynamicist), Tony Fewing (mock up), Peter Kubicki (assistant designer) and Dave Walker (flight test). A year later he was joined by Robert Page (chief technician) and Dave Walker became head of the experimental workshops, thereby giving Teddy control of development and production. These movements are summarised in the schematic of Petter's collaborators over twenty years and three companies, showing he must have inspired respect and loyalty.

W.E.W Petter's companies and collaborators (mostly recruited by himself):

Westlands (1929–44)	English Electric (1944–50)	Folland Aircraft (1951–59)
Whirlwind Welkin Lysander	Canberra P1 Lightning	Midge Gnat
Robert Bruce (Managing Director)	George Nelson (Managing Director)	Henry Folland
Eric Mensforth (Managing Director)	Arthur Sheffield (General Manager)	T Gilbertson (General Manager)
Arthur Davenport (Designer)		
Petter (Tech. Dir.)	Petter (Chief Engineer)	Petter (Managing Director)

G. Hill (Design Engineer)	Frederick Page (Asst Chief Eng.)	
	Don Crow (Chief Draughtsman)	Joe Bulger (Chief Designer)
John Digby (Asst Des. Eng)		John Digby (Special Projects)
Glen Hobday (Stress)	Glen Hobday (Martin Liaison)	
H. Harrison (Chief Draughtsman)	H. Harrison	Gordon Hudson (Chief Stressman)
Robert Page (Stress)		Robert Page (Chief Technician)
	B. Ollie Heath	
	Ray Creasey (Aero/Project Design)	
	Dai Ellis (Chief Aerodynamicist)	
	Roy Fowler (Wind Tunnels) (NPL)	Roy Fowler (Chief Aerodynamicist)
	Alan Constantine	Alan Constantine (Asst Designer)
	Tony Fewing	Tony Fewing (Mock-Up)
	Dave Walker	Dave Walker (Chief Flight & Exp. Shops)
	Peter Kubicki	Peter Kubicki (Asst Designer)

Many of the existing staff were young and without much experience beyond guided weapons. Some thrived in a team dedicated to design and manufacture, but many of the old Folland staff chose to leave rather than accept demotion and subservience to the newcomers.

Petter now had to use his ingenuity to design a lightweight fighter that would be competitive with the contemporary front-line fighters. He was not alone in trying to reduce the weight and cost of interceptors, but many designers were attempting to lighten or alter a previous design. Petter's solutions were usually unorthodox, including a proposal to the RAF for a fighter with expendable gas-turbine engines fitted to a light airframe as quickly detachable units. This would be good for maintenance, but some services described this project as a semi-expendable aircraft.

In 1951 the air force was becoming concerned about Russian Tu-4 bombers flying, and escorted, at 35,000ft. A delegation visited Petter in July to discuss a lightweight interceptor with an all-up-weight of 5,000lb and a vertical climb rate of 2.5 minutes to 35,000ft. A staff target was proposed for twenty-four aircraft for use in tactical trials. An air force staff target (AST O/R/303) was issued, stating, 'The primary and most serious threat to the UK lies in many raids by Soviet Tu-4 bombers of which large numbers are being built. By 1954 about 1,000 aircraft of this type is likely to be available to the Russians.'[3]

Although the RAF and Air Ministry were sympathetic to the need for a lightweight fighter, a major problem for Petter was the availability of suitable engines with a high thrust/weight ratio. Lord Hives of Rolls-Royce had stated that the company could no longer provide an engine that could be 'hooked on' to Petter's projected aircraft. Armstrong-Siddeley Motors (ASM) had the Viper as a possible solution, just for the prototype, and it would eventually be the power unit for the Jet-Provost trainers. Petter was not deterred by this impasse and decided to proceed with a prototype airframe powered by the ASM Viper engine – this version of the aircraft would be known as the Midge Type Fo.139. In January 1952 he forwarded to the Air Staff the Folland brochure of the Fo.140, covering two versions, one powered by the Bristol BE22 Saturn, and the other by the Rolls-Royce Derwent. The former, at an all-up-weight of 6,400lb, was for a lightweight interceptor role; the latter, at 8,000lb, was for ground attack and had a longer range. These aircraft had wings swept at 40°, a modest wing loading of 50lb/sq. ft, armament of two 30mm Aden cannons, and a variety of rocket projectiles, or alternatively two 1,000lb bombs. Two side intakes ducted air to the single engine located at the rear of the fuselage, with no reheat. The tailplane was to be mounted on the top of the fin, which was a surprise

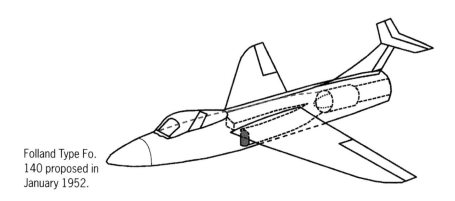

Folland Type Fo. 140 proposed in January 1952.

in view of the research carried out on the English Electric Lightning and Shorts SB5.

Having progressed through a number of design studies, Petter decided to proceed with a privately funded aerodynamic test vehicle. This was in spite of a number of Air Staff and Air Ministry officials showing little interest in his lightweight fighter project, and their not being prepared to finance any research in the development of a 5,000lb-thrust engine, which would be a crucial requirement. Some officials gave the impression that they would believe in a lightweight fighter provided it had an engine of 7,500lb thrust, weighed twice as much and was twice as large,[21] in other words another Hawker Hunter.

Petter revised his design to have higher shoulder wings and a low set tailplane. This allowed for larger air intakes moved forward, and the design was therefore able to accommodate the Aden guns instead of mounting them in the belly, and also, crucially, aligned them with the aircraft centre line. The project was now much cleaner aerodynamically and had greater space for fuel tanks or equipment. This design, designated Fo.139, became the Folland Midge, aimed to be a single demonstrator. It used hydraulically powered 'flaperons', main undercarriage doors to be used as airbrakes and a one-piece canopy that hinged over an inner armoured windscreen. Petter was pleased to find that there was expertise at Hamble in the design and manufacturing of magnesium alloy components, which he had pioneered in his previous designs. In spite of the low-powered jet engine, this little jet could break Mach 1 in a shallow dive.

During late 1952 Folland had been asked by the Ministry of Supply to undertake the development of a small, lightweight, fully automatic ejection seat. This would be ideal for the Gnat, so in conjunction with the Royal Arsenal at Woolwich, Folland developed a single slow-burning cartridge resulting in an ejection velocity of 80ft/second. It was fully automatic and usable from sea level up to 50,000ft.

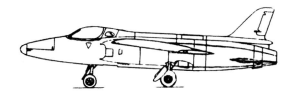

The Folland Midge.

In 1954 the team was completed with the recruitment of two test pilots. Squadron Leader Teddy Tennant DFC came directly from the RAF to become chief test pilot. He was awarded the DFC in 1944 and had flown more than 250 sorties over France, including train busting, dive bombing and rocketing sorties flying Defiants. He was joined by Squadron Leader 'Dick' Wittington DFC, who had been a production test pilot at English Electric, and had captured the London-to-Darwin speed record in a Canberra PR3. Teddy Tennant was to perform duties with outstanding skill and devotion, comparable with Petter's previous test pilots, Harald Penrose and Roland Beamont.

In July 1954 Teddy received the sad news that his father had died. Sir Ernest had returned from British Columbia several years earlier to spend his last days in his home town of Yeovil, and had finally embraced the strong religious principles of his brother Percy. Teddy and his father had still not become very close and Teddy, never one to show his emotions or discuss his private life, received the news and took refuge in concentrating on the major tasks at hand.

A target date of 31 July 1954 had been set for the completion of the Midge, and this was met. The first flight was carried out on 11 August at Boscombe Down, with Tennant at the controls. By 23 August Tennant was able to report that the Midge had flown nine hours at speeds from stall at sea level to Mach 0.95 in a gentle dive from 20,000ft. During September the Midge was displayed at the SBAC Farnborough Exhibition, and the company made the aircraft available for three days at Boscombe Down for eight pilots to make 114 flights. The Aeroplane and Armament Experimental Establishment (A&AEE) consequent report concluded with the following, 'The tests made showed the Midge aircraft, even at such an early stage of its development, to have very good aerodynamic properties. It is clear therefore that the Midge, with its excellent basic handling qualities, is a sound foundation for the more highly developed Gnat.'[21] In November three pilots from the Indian Air Force flew the Midge after Tennant had carried out some spirited flying displays in India.

The Midge had no flaps, so with a view to improve handling and to landing at a slightly shallower incidence, the ailerons were moved inboard and functioned as flap/ailerons, drooping by 20° with the lowering of the undercarriage. Boscombe Down pilots remarked on the improvement in the landing approach. The Indian Air Force, having first evaluated the Mystere IV in France, recommended that the Gnat should be considered

The Midge in 1954.

for manufacture at the Hindustani Aircraft factory. Over a period of thirteen months the Midge logged 109 flying hours by twenty pilots, but in September 1955 a Swiss pilot mistook the trim switch on the control column for the 'press to speak' button he was used to on Vampires. The aircraft crashed nose down and the pilot was killed.

The Gnat

The Midge had been an unqualified success as far as handling, take-off and landing was concerned. Many pilots enthused about its ease of flying and manoeuvrability (including those from the RAF, Indian Air Force, Canada, New Zealand, Jordon and the USAAF). An American pilot went so far as to say: 'You don't have to climb into this aircraft, just put it on.' The only feature lacking was in performance and climb rapidly to altitude, because of course there was no suitable engine. This drawback was about to be overcome.

In late 1952 Petter met Stanley Hooker, who had moved from Rolls-Royce to Bristol Engines at Filton. Hooker was keen to support him as he knew from his own calculations that it should be quite possible to design a simpler axial engine design weighing about 800lb, and giving 5,000lb thrust, which filled Petter's requirement exactly. Hooker was also personally keen on this challenge because he had hitherto spent his efforts improving other people's designs, such as Henry Royce's Merlin, Frank Whittle's W2B, and Frank Owner's Proteus and Olympus.[10] He felt that by now he had served a sufficient apprenticeship to embark on his own designs, and was eager to do just that. He saw that the Gnat offered opportunities to rethink axial-engine design and introduce a number of novel features.

One was the elimination of the centre bearing. All previous axials had three shaft bearings, one at each end and one in the middle. It was not possible in those days to keep three bearings absolutely in line, so the assembly had to be split into a front and rear section joined by a coupling able to be adjusted for small misalignments. The proposed BE 26 Orpheus adopted a short seven-stage compressor driving a single-stage turbine and no centre bearing. In a conventional engine this would mean a main drive shaft susceptible to whirling and whipping around like a skipping rope. Hooker made the shaft as a thin-walled circular tube with a diameter greater than 200mm. This gave the shaft a huge increase in torsional stiffness (Appendix 2) and raised the whirling speed way above the operational speeds. The omission of a central lubricated bearing meant a lighter supporting structure, easier to maintain. The combustion chamber was also new, consisting of a single casing containing seven flame tubes: this meant no mechanical troubles typical of the traditional assembly of steel castings and sheet metal parts. The first engine ran in 1955 and every one of its radical design innovations worked like a charm. Over the next year or two the Orpheus was type tested at 4,850lb thrust and still weighed 800lb. The Derwent and Avon engines had thrust/weight ratios of three, but the Orpheus pushed this up to six.

The Fo.141 Gnat prototype's design basically followed the Fo.140 project; the wing design was similar, with an 8 per cent thickness/chord ratio and a

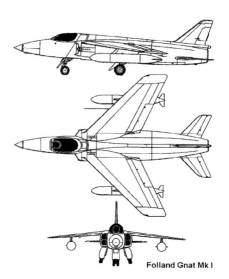

The Gnat Mk. 1.   Folland Gnat Mk I

sweepback of 40° at the quarter chord line. But the wing area was increased from 125 to 136sq. ft. Its length was increased from 29ft to 30ft, the wing span from 21ft to 22ft, and the maximum take-off weight was doubled from 4,480lb to 9,000lb. Flying at 45,000ft and cruising speed of 530mph the Gnat's fuel consumption was only 135 gallons per hour. The take-off distance to 50ft was 2,600ft and landing distance from 50ft was 3,500ft. Climb to 45,000ft took 5¼ minutes.

The primary armament on the Gnat consisted of the 30mm Aden Mk IV revolver-type cannons mounted in the outer portions of the air intake ducts, with 115 rounds per gun. On other aircraft, such as the Hunter and the Swift, the gun muzzle blast effect disturbed the airflow to the engines, causing flame-out at high altitude. Petter said to Gordon Hudson, 'You, Hudson, will design a blast suppressor to prevent flow disturbance.'[21] A row of four circular holes on the outside of each barrel and extensions of each intake in front of the muzzle kept gas from being sucked into the engine. This device was very successful and was subsequently used on both Harrier and Hawk.

The most significant change from the Midge was the recognition that the manual elevator was inadequate at transonic speeds even when backed up by the electrically actuated tailplane. One solution, used by conventional fighters, was to have a power-operated tailplane, using a hydraulic system. Petter refused to accept the weight penalty of this option on the Gnat. He decided to retain the elevator for back-up and to provide 'feel' to the pilot with a single hydraulic actuator; geared from the elevator, this would move the tailplane and provide the major longitudinal control. The electrical control of the tailplane was retained as a back-up for emergency landings without hydraulic power.

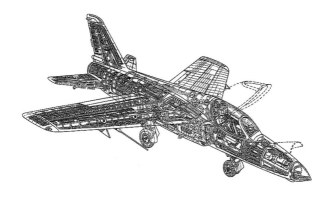

Basic structure of the Gnat Mk. 1.

Head on view with external
armament and drop tanks.
Note undercarriage covers
acting as airbrakes too.
(Courtesy of *Flight*)

The main wing was built as a complete structure passing through the top of the fuselage and could be unbolted from the fuselage frames and removed and replaced in hours. It was a simple two-spar torsion box stiffened by 'Z' section stringers parallel to the rear spar. There were no stress-raising cut-outs required for storing an undercarriage. The fuselage was a light alloy thin skin stiffened by light alloy pressed frames and extruded stringers. The fuselage could be broken, behind the main engine mounting frame, for installation and removal of the engine. The only large machined parts were the combined main undercarriage and rear wing attachment points, and the tailplane attachment/ frame castings. The main undercarriage and the nose wheel landing gear all had covers that, when they were deployed, acted as airbrakes as well.

This Gnat development had not escaped the attention of the USAAF. Its progress was followed by Colonel Johnnie Driscoll, head of the MWDP (Mutual Weapons Development Program) office in Paris, whose remit was to organise projects within NATO and bring in additional nations who had a little industrial strength. The NATO supreme commander was General Lauris Norstad, who was interested in light fighters because they could be dispersed widely away from vulnerable airfields in Europe, and also because they were within the capacity of less industrialised nations such as the rebuilt Italy and West Germany. Driscoll therefore convened a meeting of NATO aircraft companies and gave them details of a requirement for a light tactical strike fighter. It had to weigh less than 8,000lb and be able to take off and land in less than 2,000ft in what Driscoll called 'cow pastures'. At this time all jet fighters required hard runways of at least 5,000ft and were vulnerable to the new air-to-ground missiles, which were just appearing.

The requirements for the NATO strike fighter were formalised in August 1954, primarily for a tactical strike day-fighter, for a ground attack role, which would target airfields, trains and barges; the aircraft also needed to be capable of operating from semi-prepared surfaces and have a take-off distance (to clear 50ft) of 3,500ft or less. The aircraft should be very simple to maintain in the field and sufficiently robust to withstand operations at high intensity, with major components interchangeable to permit repair by replacement.

The specification was almost exactly Petter's concept for the Gnat, and so he had the ball at his feet. However, there was one feature in the specification which turned out to be crucial. Petter's concept was for a cheap, lightweight fighter that could quickly intercept high-altitude bombers and take off from conventional runways. The NATO requirement was still for a large number of cheap interceptors but widely dispersed on a large number of substandard airfields. The Gnat had thin high-pressure tyres for concrete runways. Stanley Hooker pleaded with Petter to change to low-pressure tyres, but his response was always that he could not do it because larger bays in the fuselage (which would be required to house the larger retracted wheels) would increase drag. Hooker, in desperation, urged him 'just to say that you will try to do it', but Petter, with his Quaker upbringing, could not stretch the truth in the slightest.[10] So the NATO light fighter competition ended up being between Breguet, Dassault and Nord in France and Fiat in Italy; other British firms showed no interest.

The NATO competition panel was composed of two French, two Italians and two British judges, with an Italian panel leader. The French and Italian judges decided to place the Gnat last in selection (in spite of its two-year lead), mostly because of its failure to fit low-pressure tyres, but also because its take-off distance was almost as long as the Hunter's. Petter's appearance before the panel did little to advance his case, because he was not prepared to introduce modifications to facilitate operation from grass. Robert Page said of the Gnat:

> it was a strong contender, backed by the Air Ministry, but missed out because of Petter's obstinate refusal to incorporate low pressure tyres for operation off temporary airfields, that would mean slight external blisters on the fuselage sides and possibly some reduction in the top speed. Later on in the Gnat Trainer version we got the wider tyres without any blisters.[21]

France's Sud-Est Durandel and Sud-Ouest Trident teams dropped out, as did Nord, leaving Breguet to build the Taon (French for 'firefly' and an anagram of NATO); Dassault the Etendard; and Fiat the G91. The satisfying thing for Hooker was that all three finalists chose the Orpheus as their engine, so Driscoll was able to dip into his pile of dollars and assist him to develop the engine. The deal was that the US would pay for 75 per cent if the developing country would find the rest. As Whitehall showed no interest, Hooker asked Verdon Smith at Bristol to pay for the Orpheus development. He agreed at once and hastened the programme into the manufacturing stage. It is ironic that the Orpheus was so successful and the Gnat not so at this stage, when Petter was the inspiration for Hooker to design such an engine.

Thus the Gnat was rejected, in spite of the fact that it was the only design that would be ready and flyable in 1955 and operational by 1957, the date specified by NATO, and the only one below the maximum weight limitation of 8,000lb. At the conclusion of the competition the only Italian submission, the Fiat G91, was selected, which aroused passionate French calls of foul and the refusal of the French government to commit themselves to 'that' aircraft.

The first flight of the Gnat prototype was on July 1955, with Tennant at the controls, and was most successful and uneventful. Handling was superb.

Below: Breguet TAON, AUW 11,000 lbs.

Above: Dassault Etendard , AUW 12,900 lbs. (Courtesy net Maritime)

Right: Fiat G91, AUW 10,000 lbs.

It was highly manoeuvrable and had a roll rate of more than 360° per second. This was largely due to its small polar inertia and low wing loading of 49lb/sq. ft.

The first flight had been preceded by a potential government order for six pre-production Gnat Mk 1s; by May two of these were under construction and structural testing started on the wing–fuselage joints. The progress outlined for these first six Gnats was:

1. armament trials
2. aerodynamic trials
3. engine development
4. performance trials
5. electric/electronic trials
6. evaluation by the A&AEE.

Petter proposed to fit a 6 per cent thin wing for the Gnat on the seventh, eight, ninth and tenth aircraft; the ninth and tenth would have Orpheus engines fitted with reheat. The ministry emphasised that the six Gnat aircraft were for development only and would not be used by the RAF.

Throughout this development period all the costs had been covered by company funds, and by July 1955 the expenditure was between a half and three-quarters of a million pounds. Fortunately, with the success of the Indian flights in the Midge, the Indian government had been interested in both buying and building the Gnat for its own air force. Although the Ministry of Supply had no intention of buying it for the RAF, the British government was prepared to sponsor the six prototypes to give confidence to the Indians and to assist export orders.

For the Farnborough show in September, Teddy Tennant had the first prototype ready to be demonstrated at Mach 0.8 at sea level; it had a very impressive rate of climb and manoeuvrability. Some RAF pilots were received at Hamble and flew some short flights out of Chilbolton. Their enthusiastic reports encouraged Petter to turn the Gnat over to the Central Fighter Establishment for an early evaluation. Their pilots were extremely impressed, to the extent of calling it 'an aerodynamic masterpiece' in their official report.[4] They described climbing to 45,000ft in five minutes, at which height the rate of climb was still 4,000fpm. It should be remembered that these flights occurred before the final fully powered tailplane had been installed.

Throughout his career Teddy Petter was very well served by all his test pilots, although he didn't always show great appreciation of their efforts.[3] Teddy Tennant was a remarkable man of many talents. For his home he rented a farm on the edge of the Chilbolton airfield so that he could ride his horse to work. He would relax by scything the airfield grass to provide fodder for his animals, and his evenings were often spent running a pub in Stockbridge with fellow test pilots from Vickers-Supermarine, who shared the Chilbolton airfield with Folland. In September 1956 he made a most remarkable escape from a hazardous situation brought about by a simple design flaw in the Gnat.

The standard test routine for assessing critical aero-elastic flutter speeds was for the pilot to give the stick a sudden jerk and see if the consequent oscillations died away at the chosen speed. If they did, then the test was repeated at gradually increased speeds until the damping disappeared and the onset of unstable oscillations became apparent. This procedure was particularly important for the Gnat tailplane because both the elevator and the tailplane were movable and a coupled oscillation could be a recipe for resonance and flutter. Tennant performed these tests at about 1,000ft above the airfield and all looked well up to Mach 0.85.

Tennant didn't realise that at Mach 0.91 a severe torsional flutter had caused both halves of the tailplane and the elevator to separate from the aircraft. It so happened that, at this weight, speed and altitude, the vertical lift on the tail was zero, so no disastrous pitch up and down occurred. Finding no response to stick movements, Tennant rolled the aircraft slightly to line up for an emergency landing, but on receiving no response he realised he had no option but to eject. His air speed was way above that for earlier ejector seat tests, but he pulled down the face blind and all the required automatic events took place successfully, including the parachute deployment. He made a perfect landing within walking distance of his farm, where he poured himself a stiff drink, collapsed into his armchair, and tried to figure out what had happened. His wife, Syl, returned from a shopping trip and asked, 'What are you doing here? I thought you were supposed to be working today!'[3]

The two halves of the tailplane were recovered and they showed clear evidence of torsional failure of the actuating tube connecting the tail to the fuselage. It came to light that a mistake had been made in communication between the design and stress offices. A stiffening strut, connecting the tailplane root rib leading edge to the torque lever pivot, had somehow been omitted from the workshop drawings. This episode appears to be the only significant one in Petter's career where some critical item evaded his normally dedicated personal scrutiny.

This incident caused Petter to lock and eliminate the elevator's freedom of motion, but to retain elevator deployment as a back-up should the powered tailplane's hydraulic source fail. An artificial 'feel' system to the pilot's stick was incorporated, in the form of an adjustable spring geared to the hydraulic valve and locking pin on the elevator, the spring stiffness being chosen after many flight tests.

There was still no sign that the British government would ever order Gnats for the RAF, but in the meantime the Indian government was becoming concerned that Pakistan was getting assistance from the US in acquiring F-86s, an aircraft that Folland had claimed the Gnat could out-fly and out-manoeuvre. The Ministry of Supply's approval of the six Gnats for trials was sufficient to convince the Indian Air Force and in September 1956 an order was made for twenty-five aircraft, plus fifteen more to be delivered as major assemblies. This enabled Hindustani Aircraft Ltd to use them as guidelines for initiating their own production programme, in which they would have full active participation from Folland's technical staff.

The first public display of the Gnat in India took place in January 1958 when the second aircraft completed for the Indian Air Force took part in a fly-past over New Delhi. During the spring four Folland-built Gnats were delivered for evaluation by the IAF pilots. The hydraulic-powered tailplane proved to be a learning exercise for manual reversion as there was as yet no emergency hydraulic back-up system, but HAL and Folland could be congratulated in the way that they wrung out all the problems with the Gnat as they arose. They made it into the first serviceable aircraft with external fuel tanks and the carriage of bombs on outboard wing stations. The first

Petter signing contract with Indian High Commissioner Mrs. Pandit Nehru and Sir Reginald Verdon-Smith of Bristol Engines.

HAL-built machines were finally completed in May 1962, long after Petter had left Folland but he had paid several visits to the HAL production line in Bangalore, and made friendships with many Indian pilots. His relationships with Indian government officials were not so cordial, due to arguments over permissible changes in contract costs. By 1982 the IAF had bought 200 Gnats and Ajeets (an upgraded version) to bring the total in service to 235. Some Ajeets were still in service in 1991.

Finland ordered twelve Gnats for introduction into service in 1959, bringing the total sales for the Folland plant to fifty-four, nowhere near the number needed to make a significant profit. The profit from a British government order would come from the Trainer version.

The Gnat Trainer

By 1955–56 there was a clear need for a replacement for the Vampire T11 and the Jet Provost as trainers for RAF Training Command. Pilots needed training suitable for the new fighters coming into service with a superior performance, particularly in climb rate. Petter has given credit to Air Vice Marshal Richard Atcherly for seeing a gap before pilots moved on to the Hunter T 7 trainer. Folland were given to understand that a specification for an advanced trainer was in the offing, which resulted in the firm's project department being asked to scheme a two-seater trainer based on the Gnat. Tandem seating had been advocated by the USAAF and the change from side-by-side had been accepted by the RAF.

Petter's own requirement for the design study was that a trainer should be a modest change from the Gnat fighter and with a very similar performance. To achieve this was a real challenge for the project department because adding a second seat forward in the fuselage meant moving the Orpheus back to maintain the centre of gravity position. After a bit of juggling[21] the problem was solved by Peter Nelson-Gracie (an ex-Fleet Air Arm pilot), who suggested the removal of the equipment behind the pilot's seat, including the 20 gallon fuel tank. Even so, the extra seating resulted in an extension of the fuselage length by 12in. Petter naturally agreed to 9in, and his chief technician, Robert Page, remarked, 'We should have said 18 inches, then Petter would have settled for 12 inches!'[21] Petter asked the engineer in charge of cockpit design, Tony Fewing, if anything could be done to improve the cockpit. Tony replied, 'Yes, lengthen it by another 6 inches.'[3] Eventually Teddy Tennant was asked to sit in the intended cockpit for a two-hour extended mission. After crawling out

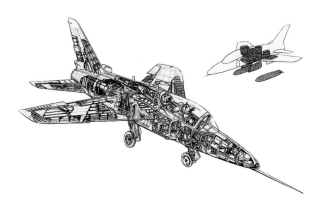

Gnat Trainer and inset showing dispersal of equipment, fuel, internal and external, behind the double cockpit. (Courtesy of *Flight*)

and uncramping his limbs, he went to Petter to insist upon the minimal length increase.

The approach speed of the fighter was too high for a trainer, and would increase still more at the higher weight. Replacing the flaperons with proper flaps made little difference so a larger wing area was needed; it was increased to 175sq. ft, compared to the fighter's 136sq. ft. To offset the consequent increase in drag, the thickness/chord ratio of the wing was reduced to 7 per cent. This, together with the inboard slotted flaps, and outboard ailerons, produced a 40 per cent increase in lift.

As the wing was going to change, Petter decided to make changes in its construction. The thinner wing would need to be multi spar (see Appendix 2), the spar attachment in the wing skin were made by chemical milling of the skins beforehand. Petter was impressed by the low cost of this new method of manufacture, and eventually the constructors, Hamble Ltd, became world leaders, carrying out much milling for Hawker Siddeley. The larger internal volume of the wing became an integral fuel tank so the aircraft would have a flight duration of 75 minutes, or 150 minutes with external drop tanks. By early 1958 the design, and some testing, had progressed far enough for the Ministry of Supply to award a contract for fourteen development aircraft. The requirements were: take-off distance, 1,600ft; landing distance from 50ft, 3,900ft; time to 40,000ft, 7 minutes; max speed at 36,000ft, Mach 0.96. This was to be Teddy's final design project to go into operational production.

The contract schedule had a first flight date of August 1959, and Teddy Tennant achieved this on the very last day. By a supreme effort, ten days' flying was completed in seven days, so it was possible to make the Gnat Trainer's public flying debut at the SBAC Farnborough show a week later.

The Gnat Trainer was of course a big jump forward for pilots coming from the Jet Provost, and it is interesting to compare the performance with other trainers, including the Hawker Siddeley Hawk, which would eventually replace the Gnat.

	Jet Provost	Gnat Trainer	Hunter T 7	Hawk
Weight (lb)	4,000	8,000	18,000	10,000
Wing area (sq. ft)	214	175	360	180
Speed at 30,000ft (mph)	460	725	725	700
Time to 45,000ft (minutes)		12	14	20
Ceiling (ft)	31,000	48,000	50,000	44,500
Wing loading (lb/sq. ft)	20	46	50	56

In spite of costing less than half the price of the Hunter Trainer, the Gnat performed as well as the Hunter, and even compared favourably with the much later Hawk. Teddy's position as chief executive officer of Folland, his vision of a radical lightweight fighter, and the technical success of the Gnat must have given him great personal satisfaction. But there were ominous signs that he would soon be overtaken by events which would deprive him of his independence.

Petter had proclaimed in 1955 that the size of his ideal aircraft firm should be small enough to be supervised completely by one chief engineer.[23]

Teddy as CEO of
Folland Aircraft Ltd.

The best unit of between one hundred and two hundred technical and design staff would have a greater degree of enthusiasm and energy in a small to medium firm than in a very large one. Petter's philosophy of avoiding large and complex aircraft came to a head in his implied and public criticism of his production director, Tom Gilbertson. The relationship between these two autocratic men gradually deteriorated, and eventually Gilbertson was dismissed and replaced by Stan Rymell, who had earlier taken over as chief inspector from Eric Holland (son of Henry), whom Petter had also dismissed. Several members of senior staff were dismissed by Petter having failed to meet his exacting standards. He wielded the axe quite ruthlessly in such circumstances.

Matters came to a head in 1959. The government had now committed itself to the policy which he had feared it would. This was the 'rationalisation' of the British aircraft industry into two consortia in the belief that fewer types of more sophisticated military aircraft would be needed by the RAF in future. The order for fourteen Gnat Trainer prototypes would be used as a stick by the ministry, who told Folland that if they wished to remain in the military aircraft business they should join the Hawker Siddeley group. Teddy realised that he would effectively be reporting to Sir Sidney Camm, with whom he had always got on well, but he felt that working under him would be impossible. The other consortia, British Aerospace, was going to be led by English Electric, who were unlikely to welcome him back.

Petter had nowhere to go, so on 11 November 1959, at the age of 51, he announced his resignation to the Hawker Siddeley board. In the preceding month the managing director of Bristol Aircraft Ltd, Peter Masefield, wrote to Petter to enquire whether rumours of his impending resignation were true, and if he could help by addressing the issue with the Prime Minister, Harold MacMillan, at a lunch he had arranged. In his reply Teddy wrote:

> I believe it will need more than a Prime Minister to put things right in our industry. One can I suppose tolerate an absence of policy; one can even forgive a wrong policy if honestly arrived at, but what I cannot stomach are the abominable methods by which one aspect of this policy, affecting this company, has been carried out. There are quite a few of my team who feel the same. However the outcome might have been worse for the shareholders, staff, and the product we have created, which is assured of continuity and I think success.[3]

Folland Gnat GT1 *Yellowjack* in a display. (Courtesy of Appingstone)

In December 1959 Teddy departed from Folland and the British aircraft industry, never to return. His departure from Hamble and the UK was undoubtedly coloured by the fact that his wife, Claude, to whom he was devoted, was showing the early symptoms of Parkinson's disease, for which there was no cure, nor indeed any amelioration.

A total of 167 Gnat Trainers were eventually manufactured, the first operational one being delivered to the Central Flying School in January 1962. The Gnat continued as the RAF's advanced trainer until 1978, when it was replaced by the Hawk because it was near to its fatigue life and not because of its performance, which was slightly better than the Hawk.

The final achievement of Teddy Petter's aircraft came in 1965 when the official RAF aerobatic team switched from the Jet Provost to the Gnat Trainer and became initially the Yellowjacks, and later the Red Arrows. The aircraft's speed and manoeuvrability made it ideal for precision flying in formation, and this team became the public face of the RAF for fourteen years. By this time Teddy had lost all interest in aircraft and probably never even knew of this honour awarded to his final aircraft design.

7

THE FINAL YEARS, 1960–1968

Petter's last six months at Folland must have been a time of distress and frustration, particularly because of the manner in which his hand had been forced in agreeing to the takeover by Hawker Siddeley. In his years at Westland and English Electric, in such times he had been known suddenly to take time off and seek solace in a religious commune, only to return to industry brimming with energy, enthusiasm and his creativity fully restored. As mentioned previously, nowadays this behaviour might be termed bipolar, a condition not uncommon in individuals with exceptional talents, usually in the fine arts, where creativity is expected. Whatever the reasons for Petter's behaviour, there is no doubt that he was a deeply religious man, which owed something to the examples set by his mother, his uncle and, later, his father. However, his behaviour this time was affected not solely by frustration at work but by the deterioration in health of his beloved wife, Claude. His next encounter with authority was due to his wife meeting a man who was to dominate both their lives.

In 1955 Claude had met a 'Father' Forget at her aunt's house in Paris.[3] He had described himself as a former minister of the Reformed Church of France who, with the help of God, could cure ailing people through communal prayer. In 1957 she met him again at her parents' house in Geneva, where Forget was holding prayer and faith-healing sessions. He was clearly persistent and paid the Petters a visit at their family home in Cundridge, where he urged Teddy and Claude to join his community. Claude would be cured, he told them, and they would all serve God in peace.

Whilst the Folland takeover was taking place in late 1959, Teddy took Claude to a villa in Cavalaire on the French Riviera, and probably met Forget again. In fact by November they were making preparations for a new way of life, and early in December Petter and Claude departed England to stay at the commune in Thonon, Haute-Savoie, taking their 14-year-old daughter, Jenni, with them.

Our judgement of the power of Father Forget, and the acceptance of his assurances by the Petters, can only be speculative. How could a brilliant, self-confident scientist, albeit a deeply religious man, be so compliant, and possibly confused? Fortunately we do have an insight into this world from a book, *My Keeper*, by Genevieve Burnod. Burnod was also duped by Forget, but later escaped and put her experiences in print. On the cover of her book, she asks, 'How can an intelligent and committed Christian couple be duped in to following a self-appointed and misguided spiritual leader? What makes loving Christian parents hand over responsibility for their children to others?'[25]

Genevieve Burnod was a schoolteacher living with her husband, Andre, and three children in Thonon, France; the family were members of the local French Reformed Church. They found the services rather dull, until one Sunday a new group joined the congregation and brought the service to life with the quality of their inspirational singing and responses. This group was Father Forget (who had been a missionary in Africa), his wife, two sons in their twenties and a daughter. This 'team' claimed to have experienced wonderful answers to prayers, but the leader was bitter towards the established church. Forget had been excluded from the church because of his attitude towards restrictions on his teaching methods, which were not deemed acceptable. The process of capturing the Burnod family, and all their worldly goods, was set in motion.

Genevieve wrote, 'When I think this over now, I know that what we really needed was to share things with people having Biblical knowledge and spiritual experience, so that we could discern whether we were dealing with a real prophetic Ministry or an imitation, or a lie from Satan.' An event followed which convinced the Burnods that Father Forget really did have faith-healing powers. Their youngest son, Eric, had developed a very painful stomach ache. Father Roger brought over his team to pray whilst he laid his hands on the principal area. 'Most importantly we were not to show our doubts by taking Eric's temperature any more. On the following morning Eric was bright and cheerful and his temperature back to normal. The pain had gone too and he joyfully went to school.'[25]

140

Coincidentally Genevieve's husband's firm ran into financial trouble, and Andre was charged with firing about half of the workforce. This brought about a state of depression in him, and he resigned. The family thereupon handed over their house to the commune. They undertook a 'training period', in which the sons were taken away from the local school and given to Genevieve to educate. The process of gradually separating the family members from each other began: they were given new names, and the influence of and their dependence on 'the father' was strengthened.

In December 1959, Genevieve wrote:

An English couple arrived in our house, who, we were told, were also called by God to join the team. Loving England as I did, and thinking it would be comforting to share all we were living with them, I was really looking forward to their stay with us, and we made them as welcome as we could. The wife had Parkinson's disease, and her husband (an important member of the staff of a British plane-making firm) had consulted dozens of doctors to try and help her. But, an ardent believer, he had been put into contact with our leader with his healing gifts, and had put all his hopes in him. These people arrived in our house for Christmas, and were never to return to England. At least the wife never did; her husband had to go back to settle everything, to sell out all they had, and bring back to Thonon all his fortune, which is great. Their youngest daughter joined them in our house, and was never to return to England either, abandoning there the very good private school which she loved. The 'messages' given to this man revealed that he too was called into full service, and that this would require a total gift of himself and of his belongings, but that God thus seeing his 'obedience' would give him fulfilment, and bless his family. No doubt that man felt sure then that his beloved wife would be healed.[25]

In September 1960 the *Daily Mail* printed the following under the headline 'Strange Exile of a Man of War':

The jet planes of the RAF will be screaming through the skies of Farnborough next week and on the eve of this annual demonstration of British air powers, I have news of a man who has given up all thoughts of war. Mr. William ('Teddy') Petter, brilliant designer of the Canberra jet bomber, has retired with his family to live the simple life high on a Swiss mountainside. And many of his friends are concerned about his welfare. It was in January that

51 year old Mr. Petter announced his decision to put the wickedness of the world behind him. 'I have strong religious views' he explained. 'It seems more important to me to live according to my principles than to make aeroplanes. People worship science, money and power, instead of God. In Switzerland I shall spend my time studying religion and contemplating.' Many wondered if Mr. Petter would put his principles into Spartan practice. Soon after his announcement he left with his Swiss wife and daughter for the mountains. An acquaintance who met him two weeks ago reported that he wears a plain habit and is living the life of a holy man. Mr Petter has gone to extraordinary lengths to keep his whereabouts a secret from all but his nearest friends.

Genevieve Burnod went on:

During the year 1961, the English family who had joined us no longer lived with us in our Thonon house, but in Switzerland in a much larger house which had been bought by the commune, mostly with the Englishman's money, but partly by ours too, since everything we possessed had been put into the commune. In 1962 we also were taken to Switzerland. We had been told it would be for three days only, but we never came back to Thonon. People now had no right to go out. Meals were taken by all members together, except for those who were confined to their rooms, or to have to eat at a separate table with his or her back to the others. Can you imagine children seeing their parents treated in this way, and having it explained that it was because of their disobedience to God?

During the summer of 1963, after one or more of my rebellions, I was locked in my room for more than 60 days without anything to do but watch the beautiful trees outside. My bible had been taken from me. I can witness to my living Lord that, if he hadn't been my comforter then, I most probably wouldn't be here to write this book. The English woman, who opposed them less violently than I, but faced their various treacheries with a kind of inertia and dull stubbornness, was, despite her severe illness, confined to her room for months. Like me, she didn't accept the breaking up of her marriage as a token of obedience to God. What I wish to stress is that people, who were perfectly normal, open hearted and generous, become real robots, annihilated, deprived of any free will, and transformed by the demonic power which was acting through that leader and his slavish team.

In 1964 Genevieve managed to escape, and on one of her final days she recalled:

> One day I was allowed to watch the bread-making. The Englishman was 'in authority' then, and he was the person who performed that very simple act. I saw him going through every phase of the procedure with such an ardent sense of the holiness of every gesture that – although I had always had a great respect for that man – I inwardly got the giggles. Nothing showed outwardly, but I did think 'poor chap.' They've convinced you that you are performing a highly religious act, and you are not conscious of it.

It seems that Teddy was then completely absorbed into the mystique of the community, even to the extent of being in charge of the Swiss operations. He was allowed to spend days in Switzerland or over the border to undertake necessary purchases. On one occasion he was tracked down by a reporter, Paul Tanfield of the *Daily Express*, who wrote an article on 'the truth behind the strangest of escape stories'. The article, 'Vow of Silence', went on: 'jet genius Petter who shuns the world for his faith, waits for a miracle'. Tanfield met Petter at the White Horse Inn, Eschallens, a village near Lausanne. Petter said to him:

> I do not care how far you have come, I am speaking to no one. This is the life I have chosen. I have strong religious views. It seems more important to me to live according to my principles than to make aeroplanes. People worship science, money, and power instead of God. In Switzerland I spend my time studying religion and contemplating.[3]

The Petter family had driven to the inn and booked three rooms, giving a Swiss bank as their home address. All three were seen only at meal times or when they went out on rare shopping expeditions. The owner of the inn said, 'We know nothing at all about them. They speak to no one.'[3]

Claude's mother, Louis Munier, living in Geneva, said:

> Claude has been terribly ill, and Teddy says he will speak to no one until she has fully recovered. Teddy has taken to a new religion which we cannot understand. He and Claude have faith that God will heal her. Until that time they will see no one and speak to no one. They live like monks.[3]

In May 1968 Teddy Petter died of bleeding from a chronic stomach ulcer condition that may have been caused by his poor lifestyle, and would probably have been amenable to normal medical treatment. He is buried in the cemetery in Béruges, near Poitou-Charentes. In the final paragraph of her book, Genevieve states that one of her sons, who was still in the commune, told her that Teddy rebelled against the leader in his final days, but the extent and substance of this final reaction is not known. Ironically Claude, whose ill health was responsible for the move to an isolated commune, survived Teddy by seven years.

It is difficult to understand how a man such as Teddy Petter, who was renowned as a tough, independent, almost autocratic person, who never suffered fools gladly, should have chosen to follow Forget for eight years. What cannot be taken from him is his reputation as a brilliant aircraft designer. Not only did he move from one site to another, but also his aircraft (apart from the Welkin) were all one-off designs, completely original in concept and unlike any others. He was a mathematical aerodynamicist, but also an instinctive structural designer. He clearly chose some very gifted teams on three separate sites, but also spent much time in the design and stress offices, talking and advising and critically assessing the engineers' efforts in designing components. I was fortunate to meet Alan Constantine, his assistant designer for thirteen years, and he confirmed that Petter was an inspiration, who helped with the development of many design workers who would never have been given a chance without his help. Teddy Petter was a man of his time, when it was still possible for a single person to design in concept a complete military aircraft. Those times are well past, and we shall never see his like again.

APPENDIX 1

BASIC AERODYNAMICS

The science of aerodynamics is continually being applied to improve the performance of aircraft, military and civil. The measures needing to be optimised include lift, drag, stability and control, take-off and landing speeds and distances. For most conventional aircraft the relevant features will be the wings, tailplane and fin, and all controls. The important choice for designers is therefore the shape of the aerofoil. Figure A1 shows how an aerofoil, at an angle of incidence α, deflects the oncoming free stream.

It is clear that the air is deflected 'downwards' and will result in an upward force on the aerofoil, and the extent of this downward momentum will depend on the curvature or 'camber' of the aerofoil, as well as the angle of incidence. However, a deeper insight into the aerodynamics is possible if we look at the nature of the streamlines as they approach and move over the aerofoil section. In the region of the leading edge the curvature of the flow over the upper surface is most pronounced and this needs a suction to persuade the flow to follow the surface. Under the aerofoil the curvature is much less so a smaller positive increase in pressure is needed. These surface

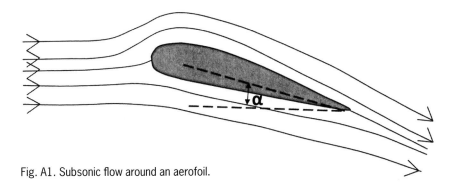

Fig. A1. Subsonic flow around an aerofoil.

pressure changes decrease towards the trailing edge. In fact the result of these distributed pressures is a force that acts roughly at the quarter chord position. This lifting force will be proportional to the applied incidence, α, but if the incidence is increased indefinitely there is a limit to the pressure gradient created and the flow will eventually break away and become detached. The wing will stall and a turbulent wake will be produced. The consequent vibrations in early aircraft were a convenient signal to the pilot in most cases. The main wing of an aircraft is designed to stall first inboard, so that the outboard ailerons are still effective for the pilot to regain control.

The resultant force on a wing section is not unfortunately in a purely vertical direction. There is a small horizontal backward component, called the induced drag. This drag is inversely proportional to the 'aspect ratio' of the wing (span divided by average chord length), which explains why gliders may have an aspect ratio in excess of thirty for instance. The optimal span-wise distribution of the lift is elliptical and was achieved directly using an elliptical plan-form in the case of the Spitfire. (In modern aircraft it is achieved by varying the effective incidence along the wing.)

In addition to the induced drag, there are three other forms: form, viscous and wave drag.

- **Form drag** is due to flow not being able to recover completely after passing over a body. If the body is streamlined then recovery can be almost complete. If the body is bluff then the flow may actually separate, leaving a base pressure on the body acting as drag.
- **Viscous drag** is due to the friction generated by the flow over the surface of the body, and depends on whether that surface is smooth or not.
- **Wave drag** is due to some flow exceeding the speed of sound.

Fig. A2. Smooth flow.

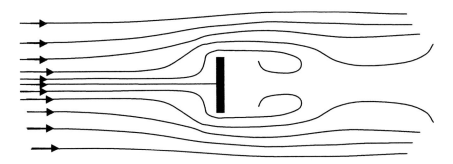

Fig. A3. Separated flow.

The higher pressures near the leading edge explain how the Lysander was able to have a leading edge slat that deployed automatically as the wing incidence was increased. The extra lift needed for take-off and landing was achieved for all early aircraft by trailing edge flaps, often simply hinged, as shown in Figure A4.

An improvement on the simple hinged flap was the Fowler flap, which moved out as well as down, thereby increasing the wing area as well as the camber. Fowler flaps later had a leading edge slat (or two) to further effectively increase the camber. More recent aircraft have controls as shown in Figure A5, where both the flaps and leading edge slats are deployed by sliding out on rollers. This not only increases the efficiency of the controls it also increases the effective area of the wing at the same time. Both controls will be deployed for take-off, but the flaps only partially because they do increase the drag. For landing the flaps have their maximum deployment because the increase in drag is now an advantage

With the advent of delta main wings, it became common to use trailing edge flaps that could also be activated differentially to function as ailerons. They became known as 'flaperons'.

Fig. A4. Simple hinged flap.

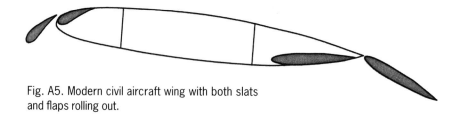

Fig. A5. Modern civil aircraft wing with both slats and flaps rolling out.

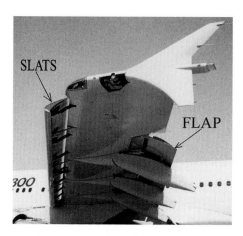

Airbus 300 Series.

Fig. A6. Devices to lesson applied control forces.

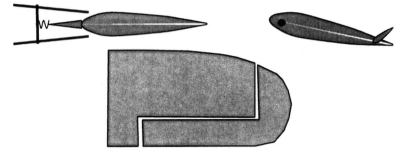

The ailerons for roll control, and the elevator for pitch, also have efficient aerofoil sections. In the period covered by this book, before powered control systems became powerful and lightweight, it was necessary to avoid asking the pilot to apply excessive forces for control. Several devices were invented to reduce control forces, and Figure A7 shows three.

The first device taps the pressure change caused by the aerofoil and applies it to a balancing lever arm ahead of the control hinge line (known as the Westland–Irving balance).

The second device shows a 'tab', which is deployed in the opposite sense to the aerofoil and results in a reduced (or zero) moment about the hinge line. A geared tab is linked to the control rods to rotate in the opposite sense to the aileron. For very large aircraft the control (for example an aileron) is completely free to rotate and the control link forces are applied directly to the tab, which requires much smaller forces to rotate. A variation on this is to insert a spring on the actuating rod, which can be 'tuned' so that the pilot feels just the right forces. This version was successfully designed by Frederick Page for his agile combat aircraft at Hawkers. The third device is a 'horn' balance, consisting of an extension of a rudder, elevator or aileron beyond the control hinge line and providing a counterbalancing moment.

The wing flow shown in Figure A1 is for subsonic flow only, thus the flow approaching the leading edge will move smoothly, either over or underneath the wing, as it receives the signal at the speed of sound. If the wing is travelling at the speed of sound this is no longer possible, and a shock wave is generated to change the flow direction instantly. This means an increase in drag, amongst other things. An aircraft may still suffer such compressibility effects even if it is still subsonic because the flow velocity on the upper surface, near the crest, will be greater than the aircraft speed. A local shock wave will be created and this may cause separation of the boundary layer as well as increased drag. To overcome this compressibility problem, three strategies were used in and after the war years:

- The wing section can be much thinner, and the leading edge sharper, as in the case of the Lockheed F104 Starfighter shown here in the West German colours.

The thickness/chord ratio of this aircraft's wing was not much greater than 3 per cent. This does make control more difficult at low speeds, and the accident rate of this aircraft was high.

Lockheed F104 Starfighter.

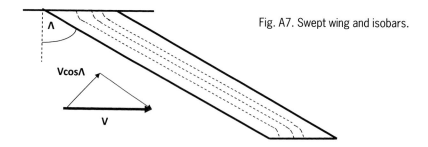

Fig. A7. Swept wing and isobars.

- The wing can be swept at an angle, Λ, as shown in Figure A7.

If the wing has a reasonable aspect ratio, the isobars will be parallel to the leading and trailing edge, so the flow may be considered two-dimensional.

The spanwise flow component along the wing has no effect on the lift. The wing lift is generated by the component normal to the wing (VcosΛ) as shown. Thus for a sweep angle of 60° the effective flow causing lift will be only a half of the aircraft speed. (The isobars shown will not be at an angle near the wing tip or root, so local changes in the wing section or plan will be used. The Gnat had a rounded tip whereas the Lightning had a change in profile near the fuselage.)

- The 'area rule'. The strength of the shocks created depends on the rate of change of the frontal area. If the cross-sectional area is plotted along the aircraft axis, then the wave drag at Mach 1 is the same as for a body of revolution of the same area distribution. This need led to some aircraft having a wasp-waisted fuselage at the section where an unswept wing joined it. The Lightning had no need of this strategy. In 1958 Ray Creasey, the deputy chief designer at English Electric, showed that the Lightning's area distribution was indeed smooth, and can avoid buffet, wing drop and other trim changes when passing through the transonic stage.[20]

Wind Tunnel Tests

To design and evaluate new aircraft designs, tests in a wind tunnel were used for all aircraft described in this book, that is in the period 1929–59.

The object of tunnel testing is to find the various forces acting on an aircraft or wing such as lift, drag, pitch and roll moments, and then to use

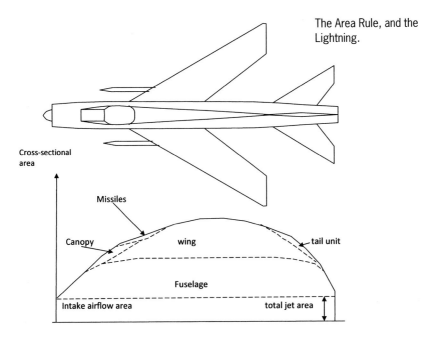

these values to calculate the behaviour of the complete aircraft at various angles of incidence, etc. The values of the various dimensional coefficients, such as lift coefficient, have to be a reliable prediction for a full-size aircraft. Whether this is so depends on the Reynolds number and how close the model/wind tunnel value will be to the real aircraft.

$$R_e = \rho VL/\mu$$

The Reynolds number, R_e, is a measure of the inertia forces to the viscous forces. The inertia forces depend on the product of the air density, ρ, the velocity, V, and a typical length scale, L, of the body. The viscous forces depend on the viscosity, μ, of the air. For models the velocity and length scale will inevitably be smaller than the real thing, so historically tunnels became larger, as did the air velocity in the working section (artificially increasing the density and decreasing the viscosity is a relatively recent achievement). For smooth flow and Reynolds number up to 2,000, the model tests will be reliable. For bigger values the flow may be turbulent, so predictions, particularly for drag and for separated flow such as large incidence and flap deployment, and the interpretation of results have to be cautious.

APPENDIX 2

BASIC AEROSTRUCTURES

The period when Petter was designing saw the end of frameworks as an aerostructure, even though they had the advantage of cheap manufacturing and battle damage tolerance, as in the case of the Hurricane and the Wellington bomber. However, the need for better aerodynamic performance meant that fabric-covered frameworks were not acceptable. We assume therefore that an aerostructure is a 'stressed skin', which can take loading in addition to keeping the correct external shape. These 'thin-walled' structures have therefore to behave like slender shallow beams in the case of the main wing, tailplane and fin. The fuselage is a much deeper stiffer structure, but also has to accommodate many cut-outs for access to equipment, engines and bombs, etc.

The nature of the internal stresses in a main wing, due to aerodynamic and inertia loading (say in meeting a strong up-gust) is shown in Figure A8. The aerodynamic lifting forces will bend the wing upwards and the inertia forces, due to the wing's and engine's masses, will bend it downwards. The internal forces at any section are labelled as the shear force, F, and the bending moment, M. This moment causes the higher stresses and can be

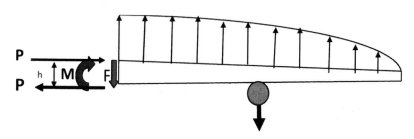

Fig. A8. Shear Force (F) and Bending Moment (M) at any section along a wing.

visualised as a pair of equal and opposite forces, P, acting on the section in the outer skin, h distance apart.

To minimise the bending stresses the lever arm, h, should be as large as possible, and the shear force requires a structure with decent vertical stiffness. One example satisfying these requirements is the ubiquitous rolled-steel joist, or 'I' beam, shown in Figure A9. It is clearly satisfying the requirement for a vertical shear web, and two horizontal flanges to cope with the bending stresses. However, this 'open' section has a very small torsional stiffness, and an aircraft wing needs to resist torsion due to both aerodynamic loading and a podded engine cantilevered forward from the wing. Poor torsional stiffness can also lead to aeroelastic instability such as divergence and flutter. Fortunately the second figure shows a typical single-cell wing box which is a closed tube and has a much greater torsional rigidity. (A closed tube of typical cross-sectional dimension, d, and thickness, t, will have a stiffness of order $(d/t)^2$ greater than an open variety.) The bending material is also placed as far apart as possible, although some early designs (such as the Lysander) relied on a simple beam forming the front spar and the leading edge to provide the closed tube.

The one disadvantage of a thin skin carrying the bending stresses is the need to prevent it buckling in compression. The buckling strain for a thin plate of thickness t and width b is proportional to $(t/b)^2$, so the effective width, b, should be as small as possible. This is achieved by adding stiffeners at pitch b as suggested in the figure. These stiffeners may be mechanically fastened to the skin or bonded. They must have a stiff section to avoid them buckling themselves, so sections such as I or Z are usual. As aircraft speeds increased and wing sections became thinner, the two stiffeners on upper and lower skins would approach each other so would be replaced by a multi-cell box, as shown in Figure A11. In addition to the need to stiffen the outer box skins to prevent buckling in compression, the vertical shear webs also need vertical stiffeners to avoid buckling in shear. (The only difference between

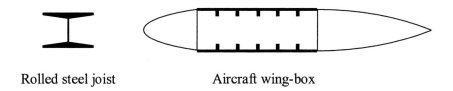

Rolled steel joist   Aircraft wing-box

Fig. A9. Open beam section and closed wing box.

Fig. A10. A multi-cell wing box.

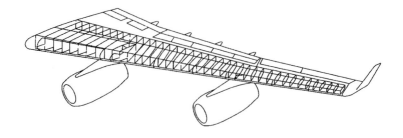

Layout of ribs in Airbus 340. (Courtesy of Airbus UK Ltd)

buckling in shear or compression is that the former buckles are at 45° to the vertical.

Even if local buckling is prevented by adequate stiffeners, a longer-wavelength buckle is possible, so internal ribs are necessary. These can be lightweight unless they have to take large loads from, say, an engine, undercarriage, or flap controls. Figure A12 shows the very large number of ribs used in the modern Airbus 340.

The fuselage is a much more respectable structure than the shallow wings, but it still needs stiffeners to stabilise its thin skin. Instead of ribs it usually has shallow frames to allow people or payload to occupy the interior. These frames will be shallow and light unless a large load is to be transferred to the fuselage from, say, a wing pick-up point, undercarriage, engine, fin or tailplane. The duty of these special open frames (or complete bulkheads) is to ensure that the loading is distributed by a tolerable shear stress around the fuselage skin.

Structural designers and stressmen need to have the analytical tools available to predict stresses in structures. Simple 'engineers theory of bending and torsion' was adequate for the aircraft described here.

Before the age of computers, stressmen had to approximate the geo-metrical nature of a structure to enable simple exact formulae to be used.

(Today an approximate computational analysis, the Finite Element Method, is used to analyse the exact structure.)

The two most-used formulae in Petter's time were the engineers' theory of bending and the Bredt-Batho theory of torsion.

For a beam with a section of 2nd moment of area I, distance of a point from the centroid y, the stress at that point due to a bending moment M, is given by:

$$\sigma = M\,y/I$$

For a closed thin-walled tube enclosing a sectional area A, having thickness t, the shear stress, τ, due to torque T is:

$$\tau = T/2At$$

For detailed stressing of joints, rivets and bolts, and compression panels the industry had assembled comprehensive data sheets based on tests. The whole aircraft, and some components, had to be experimentally tested before certification.

NOTES

1. *Plane Makers, Vol. 2: Westlands*, David Mondey (Jane's Publishing Co. Ltd, 1982)
2. *The Story of Petters Ltd*, Percival Petter (David W Edgington, 1989)
3. *Triumphs and Disasters: A Biography of Teddy Petter*, Robert Page, Roy Fowler and Adrian Page (unpublished)
4. *Adventures with Fate*, Harald Penrose (Airlife Publishing Co., 1984)
5. 'Westland Lysander', *History of War*, history of war.org/articles/weapons_westland_lysander (accessed 2012)
6. *Westland Aircraft since 1915*, Derek N. James (Putnam, 1961)
7. *Whirlwind: The Westland Whirlwind Fighter*, Victor Bingham (Airlife Publishing Ltd, 1987)
8. *The English Electric Canberra*, Roland Beamont and Arthur Read (Ian Allen Ltd, 1984)
9. *Frederick Page, 1917–2005*, Frederick Page (given to the Royal Aeronautical Society Library with conditions for it not to be published until after his death)
10. *Not Much of an Engineer*, Stanley Hooker (The Crowood Press, 1984)
11. *English Electric Canberra*, Ken Delve, Peter Green and John Clemons (Midland Counties Publications, 1992)
12. *Tender to Specification F4/40* (Westland Aircraft Ltd, 1940)
13. *The English Electric Canberra and Martin B-57*, Barry Jones (The Crowood Press, 1999)
14. *Aspects of the Design and Production of Airframes, with Particular Reference to their Coordination and to the Reduction of the Development Period*, E. Mensforth and W.E.W. Petter (Lecture to the Royal Aeronautical Society at the Institute of Mechanical Engineers, 1944)
15. *The English Electric Canberra*, Peter Bunnett (Author House, 2012)

16. *The English Electric Lightning Story*, Martin W. Bowman (The History Press, 2010)

17. *The English Electric P1 and Lightning*, Roland Beamont (Ian Allan Ltd, 1985)

18. *The English Electric BAC Lightning*, Brian Philpott (Patrick Stephens Ltd, 1984)

19. *The English Electric Lightning: Birth of a Legend*, Stewart Scott (GMS Enterprises, 2000)

20. 'Fighter Design Philosophy', R.F. Creasey, *Flight* (21 February 1958)

21. *Folland Gnat: Sabre Slayer and Red Arrow*, Victor Bingham (J & KH Publishing, 2000)

22. *Spirit of Hamble*, Derek N. James (Tempus Publishing Ltd, 2000)

23. *Where Do We Go From Here,* W.E.W. Petter (Lecture to the Institute of Production Engineers, Southampton, January 1955)

24. *Design for Production*, W.E.W. Petter (Lecture to Association Française des Ingénieurs et Techniciens de L'Aéronautique, Le Bourget, June 1953)

25. *My Keeper: When God's Love will Not Let Go: A Lesson in Spiritual Deception*, Genevieve Burnod (Marshall Pickering, 1987)

INDEX